Advances in Experimental Medicine and Biology

Neuroscience and Respiration

Volume 1047

Subseries Editor
Mieczyslaw Pokorski

More information about this series at http://www.springer.com/series/13457

Mieczyslaw Pokorski
Editor

Clinical Investigation

Springer

Editor
Mieczyslaw Pokorski
Opole Medical School
Opole, Poland

ISSN 0065-2598 ISSN 2214-8019 (electronic)
Advances in Experimental Medicine and Biology
Neuroscience and Respiration
ISBN 978-3-030-08914-6 ISBN 978-3-319-74080-5 (eBook)
https://doi.org/10.1007/978-3-319-74080-5

Printed on acid-free paper

This Springer imprint is published by the registered company Springer International Publishing AG part of Springer Nature.
The registered company address is: Gewerbestrasse 11, 6330 Cham, Switzerland

Preface

The book series Neuroscience and Respiration presents contributions by expert researchers and clinicians in the multidisciplinary areas of medical research and clinical practice. Particular attention is focused on pulmonary disorders as the respiratory tract is up front at the first line of defense for organisms against pathogens and environmental or other sources of toxic or disease-causing effects. The articles provide timely overviews of contentious issues or recent advances in the diagnosis, classification, and treatment of the entire range of diseases and disorders, both acute and chronic. The texts are thought as a merger of basic and clinical research dealing with biomedicine at both molecular and functional levels and with the interactive relationship between respiration and other neurobiological systems, such as cardiovascular function, immunogenicity, endocrinology and humoral regulation. The authors focus on modern diagnostic techniques and leading-edge therapeutic concepts, methodologies, and innovative treatments in both adults and children. Practical, data-driven options to manage patients are considered.

Body functions, including lung ventilation and its regulation, are ultimately driven by the brain. However, neuropsychological aspects of disorders are still mostly a matter of conjecture. After decades of misunderstanding and neglect, emotions have been rediscovered as a powerful modifier or even the probable cause of various somatic disorders. Today, the link between stress and health is undeniable. Scientists accept a powerful psychological connection that can directly affect our quality of life and health span. Psychological approaches, which can decrease stress, can play a major role in disease therapy.

Clinical advances stemming from molecular and biochemical research are but possible if research findings are translated into diagnostic tools, therapeutic procedures, and education, effectively reaching physicians and patients. All this cannot be achieved without a multidisciplinary, collaborative, bench-to-bedside approach involving both researchers and clinicians. The role of science in shaping medical knowledge and transforming it into practical care is undeniable.

Concerning respiratory disorders, their societal and economic burden has been on the rise worldwide, leading to disabilities and shortening of life-span. Chronic obstructive pulmonary disease (COPD) alone causes more than three million deaths globally each year.

Concerted efforts are required to improve this situation, and part of those efforts are gaining insights into the underlying mechanisms of disease and staying abreast with the latest developments in diagnosis and treatment regimens. It is hoped that the articles published in this series will assume a leading position as a source of information on interdisciplinary medical research advancements, addressing the needs of medical professionals and allied health-care workers, and become a source of reference and inspiration for future research ideas.

I would like to express my deep gratitude to Paul Roos, Tanja Koppejan, and Cynthia Kroonen of Springer Nature NL for their genuine interest in making this scientific endeavor come through and in the expert management of the production of this novel book series.

Mieczyslaw Pokorski

Contents

Advs Exp. Medicine, Biology - Neuroscience and Respiration (2018) 38: 1–6
DOI 10.1007/5584_2017_111

Published online: 20 October 2017

Prognostic Value of Osteoprotegerin and sRANKL in Bronchoalveolar Lavage Fluid of Patients with Advanced Non-small Cell Lung Cancer

W. Naumnik, I. Płońska, M. Ossolińska, J. Nikliński, and B. Naumnik

Abstract

Osteoprotegerin (OPG) is a soluble decoy receptor for receptor activator of nuclear factor-kappa B ligand (sRANKL). OPG promotes endothelial cell survival and neoangiogenesis. Dysregulation of the OPG/RANKL system has been detected in several tumors. In the present study, we evaluated the clinical usefulness of OPG and sRANKL assessment in bronchoalveolar lavage fluid (BALF) of patients with advanced non-small cell lung cancer (NSCLC). We measured the concentration of OPG and sRANKL in BALF of 44 NSCLC patients and 15 healthy volunteers taken as control subjects. The OPG content was higher in the NSCLC group than that in controls [0.48 (0.12–1.45) *vs.* 0.23 (0.14–0.75) pmol/l; $p = 0.0001$]. There were no significant differences in sRANKL content between the NSCLC and control groups [1.22 (0.74–23.00) *vs.* 1.12 (0.79–4.39) pmol/l; $p = 0.67$]. However, we found that the greater the level of sRANKL in NSCLC patients, the shorter the overall survival. We found a correlation between the content of sRANKL and the percentage of lymphocytes in BALF of NSCLC patients ($r = 0.52$; $p = 0.041$). We conclude that NSCLC patients have a higher content of OPG in BALF than healthy people. A high level of sRANKL in BALF of NSCLC patients may predict worse survival.

Keywords

Bronchoalveolar lavage fluid · Non-small cell lung cancer · Osteoprotegerin · sRANKL

W. Naumnik (✉)
First Department of Lung Diseases and Chemotherapy of Respiratory Neoplasms, Medical University of Bialystok, 14 Zurawia Street, PL 15-54 Bialystok, Poland

Department of Clinical Molecular Biology, Medical University of Bialystok, Bialystok, Poland
e-mail: wojciechnaumnik@gmail.com

I. Płońska
Municipal Hospital, Bialystok, Poland

M. Ossolińska
First Department of Lung Diseases and Chemotherapy of Respiratory Neoplasms, Medical University of Bialystok, 14 Zurawia Street, PL 15-54 Bialystok, Poland

J. Nikliński
Department of Clinical Molecular Biology, Medical University of Bialystok, Bialystok, Poland

B. Naumnik
First Department of Nephrology and Transplantation with Dialysis Unit, Medical University of Bialystok, Bialystok, Poland

1 Introduction

The cancerogenic microenvironment is composed of a complex, interconnected network of stimulators, interacting with fibroblasts, endothelial cells, immune cells, and various specific cell types depending on the location of the cancer cells (Renema et al. 2016; Travis et al. 2004). These factors interact in the autocrine/paracrine/endocrine manner involving cytokines and growth factors. Among them, osteoprotegerin (OPG) and soluble receptor activator of nuclear factor-κB ligand (sRANKL) have recently raised research interest. OPG is a multifaceted glycoprotein playing various functional roles involved in cancer sustenance and progression, such as tumor cell survival (Holen and Shipman 2006; Holen et al. 2005), resistance to apoptosis (De Toni et al. 2008), angiogenesis, and the regulation of cellular phenotype (Cross et al. 2006). OPG and RANKL expression are regulated by inflammatory cytokines released into the microenvironment of cancer cells (Gonda et al. 2009).

The receptor activator nuclear factor (RANK) is frequently expressed by cancer cells and RANKL is detected in the tumor microenvironment. Together they participate in every step in cancer development (Renema et al. 2016). Dysregulation of the OPG/RANKL system has been discovered in several tumors, such as breast (Peng et al. 2013) or prostate cancer (Chen et al. 2006). Recent studies have indicated that this system is also important in the pathogenesis and progression of non-small cell lung cancer (NSCLC) (Lu et al. 2016). There is no information on the content of OPG and sRANKL in bronchoalveolar lavage fluid (BALF) in NSCLC patients. Thus, the aim of the present study was to investigate the clinical usefulness of the assessment of OPG and sRANKL in BALF of patients with advanced NSCLC.

2 Methods

We performed this study in conformity with the Declaration of Helsinki for Human Experimentation and the protocol was approved by Bioethics Committee of Medical University of Bialystok in Poland. Written informed consent was obtained from all participant.

2.1 Patients and Control Group

Forty four 44 patients (37 men and 7 women, mean age 60 ±3 years) with a histological diagnosis of NSCLC (stages: 19 in stage IIIB, 25 in stage IV) were enrolled into the study. Adenocarcinoma was diagnosed in 8 (18%), squamous cell carcinoma in 23 (52%), and NSCLC in 13 patients (29%). The patients were subjected to routine clinical evaluation, which included chest X-ray and CT, and bronchofiberoscopy with bronchial biopsies and BALF. The fiberoscope was wedged into the segment nearest the tumor. Three 50 ml aliquots of sterile saline solution, warmed to 37 °C, were instilled into the subsegmental bronchus. Next, the fluid was gently aspirated and collected into a sterile container. The second 50 ml aliquot of recovered fluid was taken as BALF. Total and differential cell counts were analyzed in BALF samples, and the content of OPG and sRANKL were measured by an Elisa method. We used Nageotte's chamber for a total cell number, expressed as cells × 10^5/ml. The remainder of fluid was centrifuged at 800 rpm for 10 min at 4 °C. The differential cell profile was examined under light microscopy by counting at least 400 cells (at magnification 1 k). The supernatant was stored at −20 °C for further assays of protein concentration.

Clinical evaluation consisted of TNM and IASLC staging of NSCLC (Chansky et al. 2017; Woodard et al. 2016). The patients received a 21-day cycle of chemotherapy (cisplatin at a dose 30 mg/m^2 on days 1, 2 and 3, and gemcytabine at a dose 1,000 mg/m^2 on days 1 and 8 of the cycle). Four cycles of chemotherapy were given in all patients. The response to therapy was then estimated according to the response evaluation criteria in solid tumors (RECIST 1.1) criteria (Therasse et al. 2000).

The control group consisted of 15 healthy volunteers (12 men and 3 women, mean age 59 ±4 years) without any inflammatory conditions. The BALF was obtained from the right middle lobe in these subjects.

2.2 Assessment of OPG and sRANKL in BALF

Concentrations of OPG and sRANKL in the BALF supernatant were determined by the enzyme-linked immunosorbent assays (ELISA), according to the manufacturer's instructions (Biomedica Medizinprodukte; Vienna, Austria, and BioVendor – Laboratorní medicína a.s.; Brno, Czechia, respectively). The minimum detectable levels of OPG and sRANKL were 0.14 pmol/l and 0.1 pmol/l, respectively.

2.3 Statistical Analysis

The results were presented as median and minimum and maximum values. The Shapiro Wilk test was used for analysis data distribution. The respective pairs and groups were compared by a *t*-test for dependent or independent data. The Mann–Whitney U and Wilcoxon tests were used for the features inconsistent with the normal data distribution. The correlations between parameters we calculated by the Spearman rank test. Receiver- operating characteristic (ROC) curves were applied to find the cut-off level of OPG. The probability of survival and relapse-free time based on sRANKL concentrations prior to treatment was estimated using the Kaplan-Meier method. Differences were considered statistically significant at $p < 0.05$. We used Statistica v12 (StatSoft; Tulsa, OK) software for statistical analysis.

3 Results

The OPG content was higher in BALF of NSCLC patients than that in healthy subjects [0.48 (0.12–1.45) *vs.* 0.23 (0.14–0.75) pmol/l, p = 0.0001]. On the other hand, sRANKL content did not differ appreciably between the NSCLC and healthy individuals [1.22 (0.74–23.23) *vs.* 1.12 (0.79–4.39) pmol/l, p = 0.676]. These results are detailed in Figs. 1 and 2. An ROC curve demonstrates that specificity and sensitivity of OPG in NSCLC patients' BALF, relative to healthy people, was 83% and 86%, respectively, with a cut-off value of 0.324 pmol/l. The area under the curve was 0.836.

There was no appreciable association between the content of OPG or sRANKL and the stage of tumor [OPG stage IIIB *vs.* OPG stage IV: 0.48 (0.23–1.23) *vs.* 0.48 (0.12–1.45) pmol/l, p = 0.812; sRANKL stage IIIB *vs.* sRANKL

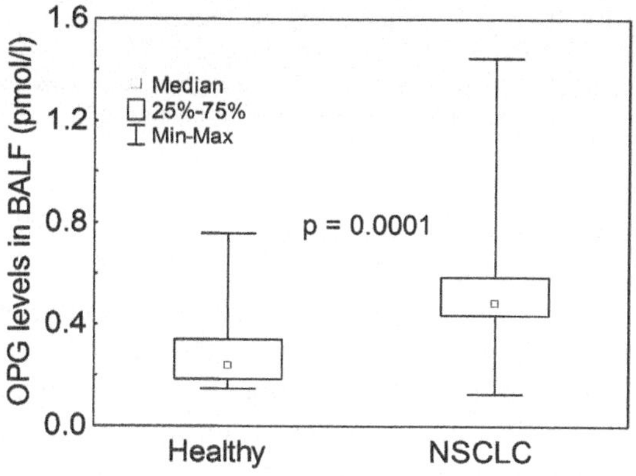

Fig. 1 Osteoprotegerin (OPG) content in bronchoalveolar lavage fluid (BALF) of non-small cell lung cancer (NSCLC) patients compared to healthy people

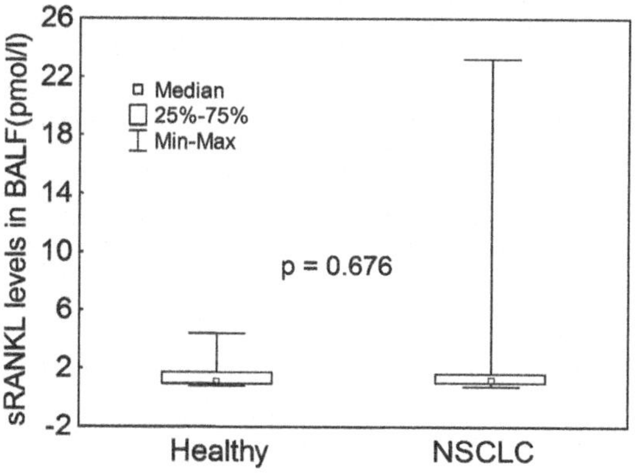

Fig. 2 Soluble receptor activator of nuclear factor-κB ligand (sRANKL) concentrations in bronchoalveolar lavage fluid (BALF) of non-small cell lung cancer (NSCLC) patients as compared to healthy people

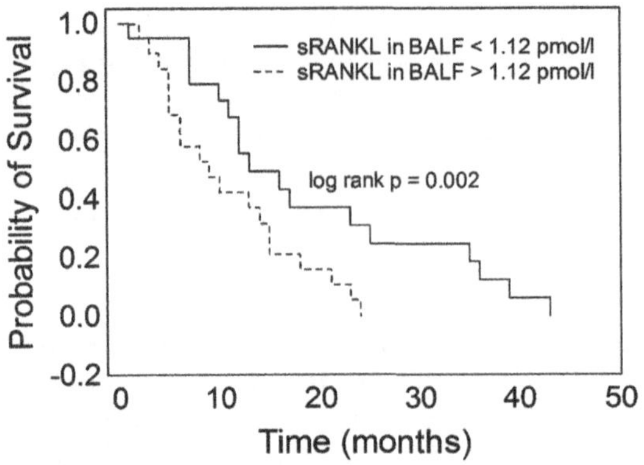

Fig. 3 Kaplan-Meier survival curve. The solid line represents survival of patients with soluble receptor activator of nuclear factor-κB ligand (sRANKL) < 1.12 pmol/l and the broken line of those with sRANKL >1.12 pmol/l. Patients with sRANKL concentration > 1.12 pmol/l had a significantly shorter survival (14 months *vs.* 17 months, respectively)

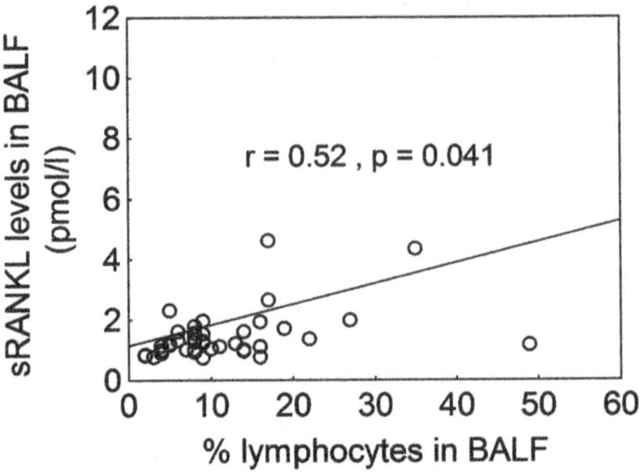

Fig. 4 Correlation between the concentration of soluble receptor activator of nuclear factor-κB ligand (sRANKL) and the percentage of lymphocytes in bronchoalveolar lavage fluid (BALF) of non-small cell lung cancer (NSCLC) patients

stage IV: 1.12 (0.74–4.6) *vs.* 1.24 (0.76–23) pmol/l, p = 0.312].

There was no significant relationship between pre-treatment concentration of OPG or sRANKL and the effect of chemotherapy. After treatment we found a partial response (PR) in 17 (39%), stabilization (SD) in 13 (29%), and progressive disease (PD) in 14 (32%) patients [OPG: PR *vs.* SD *vs.* PD – 0.51 (0.35–1.23) *vs.* 0.52 (0.32–1.44) *vs.* 0.44 (0.12–0.81) pmol/l, respectively, p = 0.198; sRANKL: PR *vs.* SD *vs.* PD – 1.24 (0.77–4.35) *vs.* 1.22 (0.98–23.23) *vs.* 1.05 (0.74–2.65) pmol/l, respectively, p = 0.225].

The mean overall survival of all patients was 15.2 ±11.0 months. We did not find a significant correlation between the level of OPG in BALF of NSCLC patients and overall survival. However, the patients with sRANKL in BALF greater than 1.12 pmol/l had a significant shorter overall survival than those with sRANKL below that level: 14 months *vs.* 17 months, p=0.002) (Fig. 3). In addition, we found a correlation between sRANKL and the percentage of lymphocytes in BALF of NSCLC patients (p = 0.041, r = 0.52) (Fig. 4).

4 Discussion

In this study we investigated the sRANKL/OPG system in BALF of NSCLC patients, which to the best of our knowledge is the first measurement of the kind. BALF is a convenient method of obtaining material directly from the tumor surroundings. The findings demonstrate a greater OPG, but not sRANKL, level in BALF of NSCLC patients than that in healthy subjects. These results are inconsistent with those of Lu et al. (2016) who failed to demonstrate a significant difference in the serum OPG level between NSCLC and healthy subjects. Further in contrast with the present findings, Lu et al. (2016) have found an enhanced level of sRANKL in NSCLC patients compared with healthy subjects. The divergent results might be explicable by advanced stages of NSCLC in our patients in contrast to those in Lu et al.'s (2016) study in which patients were in early stages of lung carcinoma before radical resection. On the other hand, the present findings are in line with the observations of Karapanagiotou et al. (2010)

who have reported that a greater OPG serum content promotes the development of lung cancer metastases. Several other studies also demonstrate that blocking OPG inhibits tumor cell proliferation (Theoleyre et al. 2004). We put forward a notion that there OPG activity may be blocked in early stages of NSCLC and becomes unblocked in advanced stages. OPG is considered a ubiquitous protein with a predominant expression in mesenchymal stem cells, immune cells (dendritic cells and T- and B-cells), and endothelial and vascular smooth muscle cells. OPG binds to glycosaminoglycans and proteoglycans, such as syndecan-1, through its heparin-binding domain, and it has a strong influence on the cancer cell development (Baud'huin et al. 2013; Theoleyre et al. 2004). In addition, OPG binds a TNF related apoptosis inducing ligand (TRAIL), a key natural pro-apoptotic and anticancer factor (Emery et al. 1998). Thus, OPG fulfils a role of an antiapoptopic factor for cancer cells (Holen et al. 2002). The present findings appear in line with the mechanism above mentioned as we noted a higher level of OPG in BALF of NSCLC patients than that in healthy subjects. These findings also are consistent with a meta-analysis performed by Zang et al. (2015) who have reported that a higher serum level of OPG may actually speak for the existence of lung cancer. We confirmed that observation in the ROC curve that clearly discriminates the diagnostic power of OPG in BALF between NSCLC patients and healthy subjects.

OPG regulates the activity of RANKL (Renema et al. 2016). RANKL is frequently detected in the tumor microenvironment, participates in every step of cancer development, and acts in a paracrine and autocrine manner on cancer cell (Roux et al. 2002). RANKL plays an important role in the immune system, increasing the ability of dendritic cells to stimulate T-cell proliferation (Lacey et al. 1998), which is, in general, in line with the present finding of a correlation between sRANKL and the percentage of lymphocytes in BALF of NSCLC patients. We also found that patients with elevated sRANKL in BALF had a significantly shorter the overall survival than those with lower levels of sRANKL. Likewise, other authors demonstrate that a high serum sRANKL level is associated with a lower survival of NSCLC patients (Lee et al. 2011). Cathomas et al. (2015) have reported on a patient with osteosarcoma in whom RANK and RANKL were expressed in the tumor cells. The patient was treated with sorafenib, with an excellent effect of a complete metabolic remission lasting for over 18 months. That observation underscores the potential therapeutic value of blocking RANK/RANKL signaling with sorafenib in osteosarcoma.

In conclusion, the assessment of OPG and sRANKL in BALF may be a useful tool in clinical evaluation of NSCLC patients. We suppose that RANKL may become an important point of future targeted therapy of NSCLC.

Conflicts of Interest The authors had no conflicts of interest to declare in relation to this article.

References

Baud'huin M, Duplomb L, Teletchea S, Lamoureux F, Ruiz-Velasco C, Maillasson M, Redini F, Heymann MF, Heymann D (2013) Osteoprotegerin: multiple partners for multiples functions. Cytokine Growth Factor Rev 24:401–409

Cathomas R, Rothermundt C, Bode B, Fuchs B, von Moos R, Schwitter M (2015) RANK ligand blockade with denosumab in combination with sorafenib in chemorefractory osteosarcoma: a possible step forward? Oncology 88:257–260

Chansky K, Detterbeck FC, Nicholson AG, Rusch VW, Vallières E, Groome P, Kennedy C, Krasnik M, Peake M, Shemanski L, Bolejack V, Crowley JJ, Asamura H, Rami-Porta R, Staging IASLC, Prognostic Factors Committee, Advisory Boards, and Participating Institutions (2017) The IASLC lung cancer staging project: external validation of the revision of the TNM stage groupings in the eighth edition of the TNM classification of lung cancer. J Thorac Oncol 12(7):1109–1121

Chen G, Sircar K, Aprikian A, Potti A, Goltzman D, Rabbani SA (2006) Expression of RANKL/RANK/OPG in primary and metastatic human prostate cancer as markers of disease stage and functional regulation. Cancer 107:289–298

Cross SS, Yang Z, Brown NJ, Balasubramanian SP, Evans CA, Woodward JK, Neville-Webbe HL, Lippitt JM, Reed MW, Coleman RE, Holen I (2006) Osteoprotegerin (OPG) – a potential new role in the regulation of endothelial cell phenotype and tumour angiogenesis? Int J Cancer 118:1901–1908

De Toni EN, Thieme SE, Herbst A, Behrens A, Stieber P, Jung A, Blum H, Göke B, Kolligs FT (2008) OPG is regulated by beta-catenin and mediates resistance to TRAIL-induced apoptosis in colon cancer. Clin Cancer Res 14:4713–4718

Emery JG, McDonnell P, Burke MB, Deen KC, Lyn S, Silverman C, Dul E, Appelbaum ER, Eichman C, DiPrinzio R, Dodds RA, James IE, Rosenberg M, Lee JC, Young PR (1998) Osteoprotegerin is a receptor for the cytotoxic ligand TRAIL. J Biol Chem 273:14363–14367

Gonda TA, Tu S, Wang TC (2009) Chronic inflammation, the tumor microenvironment and carcinogenesis. Cell Cycle 8:2005–2013

Holen I, Shipman CM (2006) Role of osteoprotegerin (OPG) in cancer. Clin Sci 110:279–291

Holen I, Croucher PI, Hamdy FC, Eaton CL (2002) Osteoprotegerin (OPG) is a survival factor for human prostate cancer cells. Cancer Res 62:1619–1623

Holen I, Cross SS, Neville-Webbe HL, Cross NA, Balasubramanian SP, Croucher PI, Evans CA, Lippitt JM, Coleman RE, Eaton CL (2005) Osteoprotegerin (OPG) expression by breast cancer cells in vitro and breast tumours in vivo—a role in tumour cell survival? Breast Cancer Res Treat 92:207–215

Karapanagiotou EM, Terpos E, Dilana KD, Alamara C, Gkiozos I, Polyzos A, Syrigos KN (2010) Serum bone turnover markers may be involved in the metastatic potential of lung cancer patients. Med Oncol 27:332–338

Lacey DL, Timms E, Tan HL, Kelley MJ, Dunstan CR, Burgess T, Elliott R, Colombero A, Elliott G, Scully S, Hsu H, Sullivan J, Hawkins N, Davy E, Capparelli C, Eli A, Qian YX, Kaufman S, Sarosi I, Shalhoub V, Senaldi G, Guo J, Delaney J, Boyle WJ (1998) Osteoprotegerin ligand is a cytokine that regulates osteoclast differentiation and activation. Cell 93:165–176

Lee JA, Jung JS, Kim DH, Lim JS, Kim MS, Kong CB, Song WS, Cho WH, Jeon DG, Lee SY, Koh JS (2011) RANKL expression is related to treatment outcome of patients with localized, high-grade osteosarcoma. Pediatr Blood Cancer 56:738–743

Lu C, Sun C, Jin H (2016) Serum sRANKL and sRANKL/OPG ratio: novel biomarkers in non-small cell lung cancer. Oncol Lett 11:2261–2265

Peng X, Guo W, Ren T, Lou Z, Lu X, Zhang S, Lu Q, Sun Y (2013) Differential expression of the RANKL/RANK/OPG system is associated with bone metastasis in human non-small cell lung cancer. PLoS One 8:58361

Renema N, Navet B, Heymann MF, Lezot F, Heymann D (2016) RANK-RANKL signalling in cancer. Biosci Rep 36(4). https://doi.org/10.1042/BSR20160150

Roux S, Amazit L, Meduri G, Guiochon-Mantel A, Milgrom E, Mariette X (2002) RANK (receptor activator of nuclear factor kappa B) and RANK ligand are expressed in giant cell tumors of bone. Am J Clin Pathol 117:210–216

Theoleyre S, Wittrant Y, KwanTat S, Fortun Y, Redini F, Heymann DT (2004) The molecular triad OPG/RANK/RANKL: involvement in the orchestration of pathophysiological bone remodeling. Cytokine Growth Factor Rev 15:457–475

Therasse P, Arbuck SG, Eisenhauer EA, Wanders J, Kaplan RS, Rubinstein L, Verweij J, Van Glabbeke M, van Oosterom AT, Christian MC, Gwyther SG (2000) New guidelines to evaluate the response to treatment in solid tumors. J Natl Cancer Inst 92:205–216

Travis WD, Brambilla E, Müller-Hermelink HK, Harris CC (2004) Pathology and genetics of tumours of the lung, pleura, thymus and heart, World Health Organization classification of tumours. ARC Press, Lyon, pp 10–33

Woodard GA, Jones KD, Jablons DM (2016) Lung cancer staging and prognosis. Cancer Treat Res 170:47–75

Zang L, Ma M, Hu J, Qiu H, Huang B, Chu T (2015) The effects of lung and prostate cancer bone metastasis on serum osteoprotegerin levels: a meta-analysis. Sci Rep 16:18324

Advs Exp. Medicine, Biology - Neuroscience and Respiration (2018) 38: 7–19
DOI 10.1007/5584_2017_128

Published online: 19 November 2017

Melatonin and Metformin Diminish Oxidative Stress in Heart Tissue in a Rat Model of High Fat Diet and Mammary Carcinogenesis

Natalia Kurhaluk, Bianka Bojkova, Marek Radkowski, Olga V. Zaitseva, Svitlana Kyriienko, Urszula Demkow, and Pawel J. Winklewski

Abstract

The aim of this study was to determine the effects of long-term administration of the oral antidiabetic metformin or the pineal hormone melatonin, and a combination thereof, in preventing oxidative stress in the heart tissue of female Sprague-Dawley rats with mammary tumors induced by N-methyl-N-nitrosourea (NMU) (50 mg/kg) given on the 42nd postnatal day. Metformin and melatonin were administered 12 days before and 16 weeks after the carcinogen. During the experiment, all animals were fed a high fat diet (10% total fat, 2.5% from lard, and 7.5% from palm oil). The findings are that mammary carcinogenesis generated oxidative stress. Reactive oxygen species (ROS) content, estimated from thiobarbituric acid reactive substances (TBARS), oxidatively modified protein content (aldehyde and ketone derivatives), and the activity of the antioxidant enzymes superoxide dismutase, glutathione reductase, and glutathione peroxidase were all augmented. Metformin caused a decrease in oxidative stress in the heart, accompanied by a decrease in diene conjugates, the elimination of ROS (stable

N. Kurhaluk
Department of Zoology and Animal Physiology, Institute of Biology and Environment Protection, Faculty of Science, Pomeranian University of Slupsk, Słupsk, Poland

B. Bojkova
Department of Animal Physiology, Institute of Biology and Ecology, Pavol Jozef Šafárik University, Košice, Slovakia

M. Radkowski
Department of Immunopathology of Infectious and Parasitic Diseases, Medical University of Warsaw, Warsaw, Poland

O.V. Zaitseva
Palladin Institute of Biochemistry, National Academy of Sciences of Ukraine, Kyiv, Ukraine

S. Kyriienko
National State University in Chernihiv, Chernihiv, Ukraine

U. Demkow
Department of Laboratory Diagnostics and Clinical Immunology of Developmental Age, Medical University of Warsaw, Warsaw, Poland

P.J. Winklewski (✉)
Department of Human Physiology, Medical University of Gdansk, Gdansk, Poland

Faculty of Health Sciences, Pomeranian University of Slupsk, Slupsk, Poland
e-mail: pawelwinklewski@wp.pl

total antioxidant status), and the activation of catalase and glutathione reductase. Melatonin caused an increase in total antioxidant status and a substantial reduction in ROS as estimated from aldehyde and ketone derivatives, lipid peroxidation at the initial (diene conjugates) and terminal stages (TBARS), and increased catalase and glutathione peroxidase activities. Metformin and melatonin combined reversed the effects of NMU on oxidative stress. In conclusion, melatonin reduces the level of oxidative stress in the heart tissue, caused by NMU carcinogenesis and a high fat diet, significantly stronger than metformin.

Keywords

Carcinogenesis · Melatonin · Metformin · Heart · High fat diet · Oxidative stress · Mammary

1 Introduction

The advancement of civilization, including the fast pace of life and profound changes in nutrition, has lead to serious health problems in individuals and the entire population. Lifestyle modifications have led to an increase in the prevalence of obesity and related diseases. Obesity is described as a chronic inflammatory process occurring in the body with the participation of visceral adipose tissue (Popko et al. 2010). It is highly correlated with many diseases such as diabetes mellitus type 2, hypertension and metabolic syndrome (Winklewski et al. 2015; Goodwin et al. 2012; Goodwin and Stambolic 2011). Factors related to lifestyle such as nighttime work, insomnia, poor diet, and a high dietary fat content, are associated with increased risk of malignant transformation. The potential mechanisms responsible for increased production of toxic metabolites and reactive oxygen species (ROS), resulting in neoplasm development, constitute an important component of basic research.

Neoplasms accounted for 7 million deaths in 2000, expected to reach approximately 10 million deaths in 2020. In Poland, there were 130,000 new cases of cancer and 85,000 deaths from cancer in 2000, and 160,000 new cases of cancer and 100,000 deaths attributable to cancer in 2010 (Syczewska-Weber and Rucinski 2008). Cancer is responsible for 25% of all causes of deaths. One of the most common forms of female malignancy in Europe is breast cancer (464,000 cases, 13.5% of all cancer cases) (Ferlay et al. 2013). In Poland, approximately 12,000 new cases of breast cancer are reported each year (Didkowska et al. 2007).

Several studies support the concept that ROS are involved in the etiology and progression of breast cancer. Importantly, high levels of the biomarkers of oxidative stress, including lipid peroxidation products, malondialdehyde or isoprostanes, protein oxidation products (carbonyls and diene conjugates), and DNA modifications are frequently identified in breast cancer patients (Kedzierska et al. 2012). It has recently been shown that altered redox status is directly correlated to estrogen levels. ROS may also control antioxidant gene expression (Sekkin et al. 2015; Carpentieri et al. 2012). Oxidative stress participates in the structural modification of estrogen and progesterone receptors in patients with endocrine-responsive breast cancer (Panis et al. 2012, Sosa et al. 2013). A high level of ROS in cancer cells may lead to a variety of biological responses, such as cell adaptation, the development of DNA mutations and genetic instability, increased proliferation rate, and resistance to some drugs used in anticancer therapy (Jou et al. 2007; Tas et al. 2005).

There are reports demonstrating that cancer patients often have symptoms of metabolic syndrome, a component of which is glucose intolerance and diabetes mellitus type 2. Therefore, an evaluation of oral antidiabetic preparations for anti-tumor properties is particularly important. Oral antidiabetics with pleiotropic properties that show anti-tumor activity have drawn considerable attention, as the incidence of diabetes mellitus type 2 and cancer continue to rise. So far, anti-tumor properties have been reported in two groups of oral antidiabetics: biguanides, particularly metformin (MF) (Erdem et al. 2015; Zhu et al. 2015; Anisimov 2014; Leone et al. 2014; Pollak 2013) and thiazolidinediones (Bojková et al. 2014).

Melatonin (MEL), N-acetyl-5-methoxytryptamine, a hormone produced primarily by the pineal gland as indolamine and secreted in a circadian pattern, protects against cancer development (Viswanathan and Schernhammer 2009), functioning as an antioxidant (Reiter et al. 2003), with properties that reduce oxidative stress (Kurhaluk et al. 2017; Esrefoglu et al. 2012) and inflammation (Ren et al. 2015; Shin et al. 2015). MEL also stabilizes mitochondrial function (Carrasco et al. 2015), protects DNA against apoptosis (Jou et al. 2007), attenuates metabolic disorders in the streptozotocin-induced diabetes model in rats (Sudnikovich et al. 2007), and is involved in the regulation of inflammatory cell infiltration and obesity-induced adipokine alteration (Favero et al. 2015).

We have recently reported a preventive or curative effect of MEL on mammary carcinogenesis, consisting of increased survival time, preferably when MEL is given in combination with other oncostatic substance (Kubatka et al. 2014; Orendáš et al. 2014; Orendáš et al. 2009; Kubatka et al. 2002). The aim of the present study was to evaluate the effect of the oral antidiabetic metformin and the pineal hormone melatonin administered alone and in combination on oxidative stress and antioxidant enzyme activity in the heart tissue in female Sprague-Dawley rats fed a high fat diet and subjected to mammary tumor cell proliferation *in vivo* induced by N-methyl-N-nitrosourea (NMU) administration.

2 Methods

The experiments were conducted following the guidelines of the European Union Council and the current laws of the Slovak Republic for animal experiments. The study was approved by the Ethics Committee of the Pavol Jozef Šafárik University and by the State Veterinary and Food Administration in Slovakia (no. Ro-2054/13–221). The animals were treated and sacrificed in a humane manner according to the principles provided in Laws No. 289/2003, 489/2003, and 23/2009 of the Slovak Republic on the care and use of laboratory animals.

2.1 Animals and Experimental Design

Mammary Carcinogenesis Model Female rats of the Sprague-Dawley strain (Velaz; Prague, Czech Republic) aged 30 days were used in the experiment. This strain is most commonly used in *in vivo* mammary carcinogenesis models, as other strains (e.g., Wistar) are not sensitive to mammary tumor induction by chemocarcinogens. The animals were adapted to standard vivarium conditions with temperature of 23 ± 2 °C, relative humidity of 60–70%, and an artificial light-dark cycle of 12:12 h (lights on from 7 a.m., light intensity 150 lx/cage). During the experiment, the animals (6 *per* cage) were fed *ad libitum* with a high fat diet (10% of total fat, 2.5% from lard, 7.5% from palm oil; Biofer, Slovakia) and tap water, or received MEL solution.

Mammary carcinogenesis was induced by N-methyl-N-nitrosourea (NMU, cat. no. N4766; Sigma-Aldrich; Deisenhofen, Germany) administered intraperitoneally in a dose of 50 mg/kg on the 42nd postnatal day. The NMU solution was freshly prepared before administration by dissolving NMU in 0.9% NaCl. A single volume of injectant was 0.5 mL.

Experimental Groups Chemoprevention with MF and MEL was initiated 12 days prior to carcinogen application and lasted until the termination of the experiment. MF (Actos; Lilly, Alcobendas, Spain) was administered in the diet at a concentration of 2000 ppm. MEL (cat. no. M5250, Sigma-Aldrich Deisenhofen, Germany) was administered in tap water at a concentration of 20 mg/L daily from 3 p.m. to 8 a.m. Only pure water was given at other times.

Animals were assigned randomly to one of five experimental groups (18 rats per group), except the control group (Group 1), having ten rats. Group 2 was NMU model group without chemoprevention, Group 3 – NMU + MF chemoprevention, Group 4 was NMU + MEL chemoprevention, and Group 5 was NMU + MF + MEL chemoprevention combined.

All rats, including control animals, used in this experiment were fed a high fat diet. All rats were weighed weekly during the experiment and palpated to register the presence, number, location, and size of each palpable tumour. NMU rats were given only the carcinogen and the values of NMU rats were compared to the control group to determine the effects of the carcinogen. The groups given the chemopreventive treatment were compared to the NMU group and to each other, to determine the effects of chemoprevention.

Food and water intake over a 24-h period was monitored during the 4th, 9th, and 14th week of the experiment (dated from NMU administration). During the experiment, the average daily intake of MF ranged from 28 to 32 mg/rat/day, which equals to *ca* 880 mg/m^2, considering the body surface of 0.034 m^2 in a 245 g rat. For comparison, the recommended daily dose of MF in diabetic patients is 1700 mg, which equals to *ca* 940 mg/m^2, considering the body surface of 1.8 m^2.

The average daily intake of MEL ranged from 0.36 to 0.48 mg/rat/day. The MEL dose was supraphysiological, about fourfold higher than the usual dose used for the treatment of sleep disorders and jet lag disorder in humans, according to the body surface area. In the last week of the experiment (week 16), the animals were killed by a quick decapitation and mammary tumors, and selected organs and tissues were removed for further evaluation.

Drugs and Solutions EDTA, HEPES, KCl, K_2CO_3, KH_2PO_4, EDTA, 2-thiobarbituric acid were purchased from Sigma-Aldrich (Deisenhofen, Germany). All drugs were freshly prepared. All reagents used in the study were of analytical grade.

Tissue Isolation The heart tissues were removed. Briefly, the heart was excised, weighed, washed in ice-cold buffer, and minced. The minced tissue was rinsed with cold isolation buffer to remove blood and was homogenized on ice in a glass Potter-Elvehjem homogenizer (Kennesaw, GA) with a motor-driven Teflon pestle. The isolation buffer consisted of 120 mM KCl, 2 mM K_2CO_3, 10 mM HEPES, and 1 mM EDTA. The pH was adjusted to 7.2 with KOH.

Heart homogenates were used for the determination of thiobarbituric acid reactive substances (TBARS) and oxidative modified protein (OMP) levels, total antioxidant status (TAS), as well as catalase (CAT) and superoxide dismutase (SOD) (1:1000), glutathione reductase (GR) and glutathione peroxidase (GPx) (1:20) activities. For the quantification of proteins, the Bradford method with bovine serum albumin as a standard was used. Absorbance was recorded at 595 nm.

2.2 Biochemical Assays

Conjugated Dienes The level of conjugated dienes was determined according to Kamyshnikov's (2004) method. Conjugated dienes are formed during the lipid peroxidation process as a result of the reconfiguration of double bonds after the detachment of hydrogen from the polyunsatured fatty acid chain. Conjugated dienes were determined by absorption at a wavelength of 233 nm and expressed in nmol *per* mg protein.

Thiobarbituric Acid Reactive Substances (TBARS) TBARS also were estimated using according to Kamyshnikov's (2004) method and were expressed in μmol of malondialdehyde (MDA) *per* mg protein from heart tissue.

Protein Carbonyl Derivatives The OMP rate was estimated using the reaction detecting carbonyl derivatives of amino acids with 2,4-dinitrophenyl hydrazine (DNFH) according to Levine et al.'s (1990) method in Dubinina et al.'s (1995) modification. The final solution was centrifuged to remove any insoluble material. The carbonyl content was calculated from the absorbance measurements at 370 nm and 430 nm with an absorption coefficient of 22,000 $M^{-1} \cdot cm^{-1}$. Carbonyl groups were determined spectrophotometrically at 370 nm (aldehyde derivatives (AD); OMP_{370}) and 430 nm (ketone derivatives (KD); OMP_{430}), and expressed as nmol *per* mg protein.

Total Antioxidant Status (TAS) The TAS level in the plasma and liver was estimated by measuring the TBARS level according the oxidation of Tween 80. This level was determined calorimetrically according to Galaktionova et al.'s (1998) method based on the reaction of Fe^{2+}/ascorbate-induced oxidation of Tween80, resulting in a stable level of 2-thiobarbituric acid reactive substances. The maximum absorbance was measured at 532 nm. The absorbance of the blank was defined as 100%. The percentage of TAS in a sample was calculated in reference to the absorbance of the blank.

Superoxide Dismutase Activity Superoxide dismutase (SOD, E.C. 1.15.1.1) activity in the supernatant was determined according to Kostiuk et al.'s (1990) method. SOD activity was assessed by its ability to produce superoxide dismutase during quercetin auto-oxidation in an alkaline medium (pH 10.0). The absorbance at 406 nm was measured immediately and after 20 min and the activity was expressed in units of SOD *per* mg protein.

Catalase Activity Assay Catalase (CAT, E.C. 1.11.1.6) activity was determined by measuring a decrease in H_2O_2 in the reaction mixture using Koroliuk et al.'s (1988) method. One unit of CAT activity was defined as the amount of enzyme required for decomposition of 1 μmol H_2O_2 per min *per* mg protein.

Glutathione Reductase Activity Glutathione reductase (GR, E.C. 1.6.4.2) activity in the blood and tissues was measured according to Glatzle et al.'s (1974) method. The enzymatic activity was assayed spectrophotometrically by measuring NADPH consumption. A blank without NADPH was used and the GR activity is expressed in nmol NADPH *per* mg protein.

Glutathione Peroxidase Activity Glutathione peroxidase (GPx, EC 1.11.1.9) activity was determined by the detection of non-enzymatic utilization of reduced glutathione (GSH) as the reacting substrate at 412 nm after incubation with 5,5-dithio-bis-2-nitrobenzoic acid (DTNB) according to Moin's (1986) method. GPx activity was expressed in nmol GSH *per* mg protein.
For protein quantification, Bradford's (1976) method was used with bovine serum albumin as the standard. The absorbance was recorded at 595 nm. All enzymatic assays were carried out in duplicate at 22 ± 0.5 °C using a Specol 11 spectrophotometer (Carl Zeiss Jena, Germany).

2.3 Statistical Evaluation

Results were expressed as means ±SE. All variables were tested for a normal distribution using Kolmogorov-Smirnov and Lilliefors' tests ($p > 0.05$). Homogeneity of variance was checked using Levene's test. The significance of differences in the level of conjugated dienes, lipid peroxidation, amino acid carbonyl derivatives, and antioxidant enzyme activities between the control and experimental groups were examined using Student's test and one-way analysis of variance (ANOVA), with *post-h*oc Bonferonni's test. Differences were considered significant at $p < 0.05$. In addition, associations between individual data were evaluated using Pearson's correlation analysis.

All statistical calculations were performed with Statistica v8.0 software (StatSoft Inc., Cracow, Poland).

3 Results

Since the process of lipid peroxidation (LPO) occurs in multiple stages, we decided to analyze the process at its beginning and end. Diene conjugation products, as the first stage of LPO primary product formation, are presented in Fig. 1. The model of mammary carcinogenesis formation induced by NMU in the rats given a high fat diet caused statistically significant changes in the substrate accumulation in the heart during the first stage of free radical production. MF or MEL chemoprevention alone after NMU administration decreased the level of LPO intensity compared to NMU rats ($F = 22.51$, $p = 0.0001$). A reduction in LPO, particularly in the content of conjugated dienes, during MEL chemoprevention was greater than that during MF chemoprevention. MF and MEL combined did not cause significant changes in tissue compared to the MF + NMU-treated and MEL + NMU-treated rats.

TBARS, estimated by the MDA level, are the end product of the terminal stages of free radical production, and are a biomarker of oxidative stress. The TBARS content increased significantly in the heart tissue after NMU treatment ($F = 5.45$, $p = 0.001$) and decreased significantly after MEL chemoprevention (Fig. 2). The intensity of the LPO process also was lower after MF and MEL chemoprevention combined ($F = 4.61$, $p = 0.002$) compared to MF + NMU-treated rats.

We further found significant effects of NMU administration on the level of OMP-AD and OMP-KD derivatives in the heart tissue and then after MF and MEL chemoprevention. MEL chemoprevention substantially limited the ROS processes compared to NMU-treated rats (Table 1).

The TAS level was statistically lower in the heart tissue after NMU treatment compared to that found in the control rats (Fig. 3). MF or MEL chemoprevention alone increased TAS compared to the NMU group ($F = 21.34$, $p = 0.0001$). Percentagewise, the effect of MEL was greater than that in both MF and NMU groups.

Antioxidant enzyme activity is shown in Table 2. NMU treatment increased SOD and GR activity, while it diminished CAT and GPx activity. MF chemoprevention augmented GR, CAT, and GPx activities compared to NMU-treated rats. MEL chemoprevention elevated GPx, and reversed the effects of NMU on SOD and GR activities. Combined chemoprevention with MF + MEL in the NMU model increased SOD activity compared to the MEL + NMU group. GR activity was lower in MF + MEL rats compared with the MEL + NMU rats.

The following associations were present in the heart tissue in the NMU model of carcinogenesis: MDA-TAS ($r = 0.89$, $p = 0.001$), AMP AD-TAS ($r = -0.91$, $p = 0.000$), and diene conjugates-CAT ($r = -0.93$, $p = 0.0001$). The association of oxidative stress biomarkers after NMU with MF chemoprevention included diene conjugates-TAS ($r = 0.88$, $p = 0.004$) and TAS-GR ($r = 0.89$, $p = 0.001$) and with MEL chemoprevention included TAS-MDA ($r = -0.92$, $p = 0.0001$), diene conjugates-OMP AD ($r = 0.87$, $p = 0.001$), and OMP AD-TAS ($r = -0.89$, $p = 0.001$). No changes in food and water intake were found in comparison with the control (intact) group. Metformin and melatonin had no significant effect on parameters of mammary tumour growth (data not shown).

4 Discussion

The present study was designed to investigate the effects of the oral antidiabetic metformin and the pineal hormone melatonin, administered alone and in combination on oxidative stress and antioxidant enzyme activity in the heart tissue of female Sprague-Dawley rats. The rats were kept on a high fat diet and were subjected to the mammary carcinogenic process induced by N-methyl-N-nitrosourea. The main finding was that metformin and melatonin, in all combinations used, prevented NMU-induced toxicity and oxidative stress. A detailed comparative analysis demonstrates that melatonin was superior to metformin as antioxidant prevention in the heart tissue. Melatonin increased a total antioxidant capacity through a substantial reduction in ROS generation and in

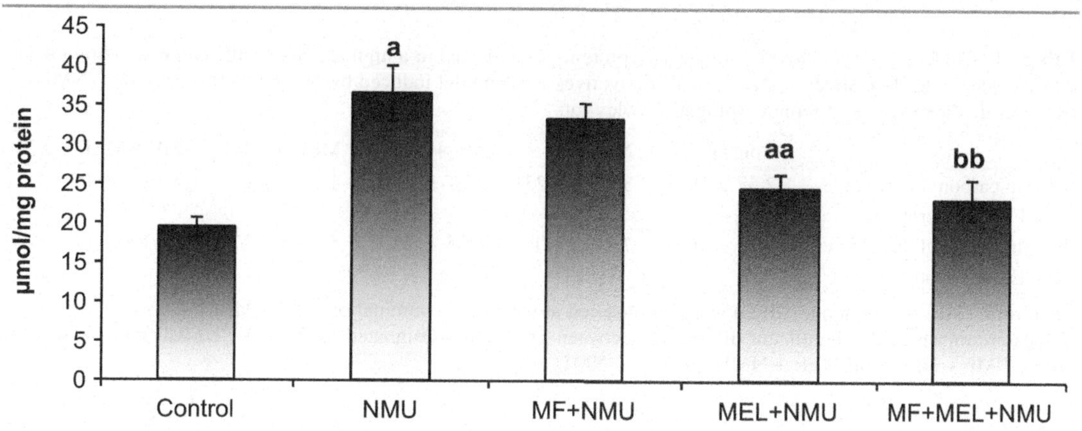

Fig. 1 Conjugated dienes (E233) in the heart tissue of female Sprague-Dawley rats subjected to a high fat diet in the mammary carcinogenesis model induced by N-methyl-N-nitrosourea (NMU). Metformin (MF) or melatonin (MEL) was administered alone and in combination (MF + MEL). Values are means ±SE. Significant differences between groups were designated as follows: **a** – NMU group *vs.* control group; **aa** – MF + NMU and MEL + NMU groups *vs.* NMU group; **bb** – MF + MEL + NMU group *vs.* MF + NMU and MEL + NMU group

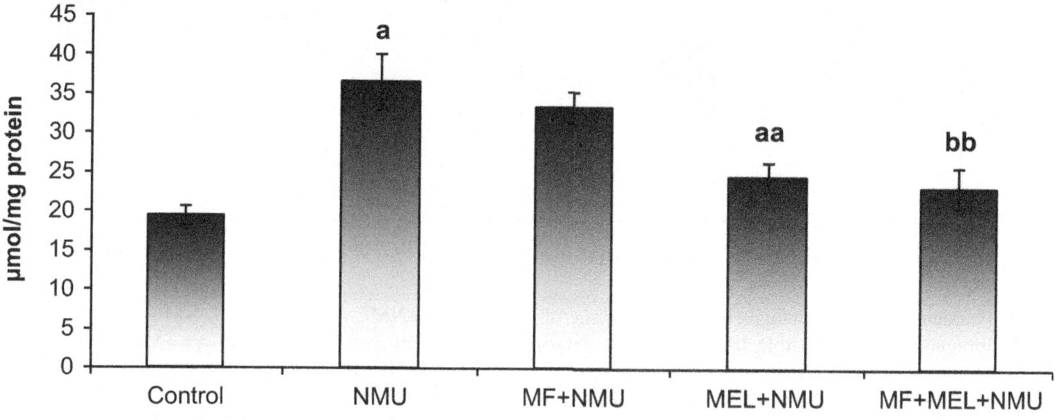

Fig. 2 TBARS products in the heart tissue of female Sprague-Dawley rats subjected to a high fat diet in the mammary carcinogenesis model induced by N-methyl-N-nitrosourea (NMU). Metformin (MF) or melatonin (MEL) was administered alone and in combination (MF + MEL). Values are means ±SE. **a** – NMU group *vs.* control group. Significant differences between groups were designated as follows: **aa** – MF + NMU and MEL + NMU groups *vs.* NMU group; **bb** – MF + MEL + NMU group *vs.* MF + NMU and MEL + NMU group

the content of oxidatively modified proteins. The antioxidant effects consisted of reductions in lipid peroxidation processes at the initial (diene conjugation) and terminal stages (TBARS estimated from MDA), protein destruction (estimated from OMP AD and OMP KD), and from the enhancement of the total antioxidant status in the heart tissue estimated from the antioxidant enzyme activity (SOD, CAT, GR, and GPx).

Data referring to free radicals and cell homeostasis suggest a coordinated activity between the enzymatic and non-enzymatic systems of ROS generation, and their removal. The literature shows a sequence of cell events in cancer cells, related to ROS generation (Tas et al. 2005). Oxidative stress promotes tumor development, but it can also be useful in the search for new therapeutic strategies for cancer treatment (Reiter et al.

Table 1 Content of oxidatively modified proteins (OMP) (aldehyde derivatives (AD); ketone derivatives (KD)) in the heart tissue of female Sprague-Dawley rats subjected to a high fat diet in the mammary carcinogenesis model induced by N-methyl-N-nitrosourea (NMU)

	Control	NMU	MF + NMU	MEL + NMU	MF + MEL + NMU
Protein carbonyl derivatives. AD, E370·mg protein	2.57 ± 0.76	8.15 ± 1.23[a]	7.51 ± 2.22	4.89 ± 1.22[aa]	5.55 ± 1.45
Protein carbonyl derivatives. KD, E420·mg protein	4.11 ± 0.89	7.44 ± 1.04[a]	8.35 ± 2.44	5.06 ± 0.57[aa]	6.33 ± 2.14

Metformin (MF) or melatonin (MEL) was administered alone and in combination (MF + MEL)
Values are means ±SE. Significant differences between groups are designated as follows: [a]NMU group *vs.* Control group; [aa]MF + NMU and MEL + NMU group *vs.* NMU group

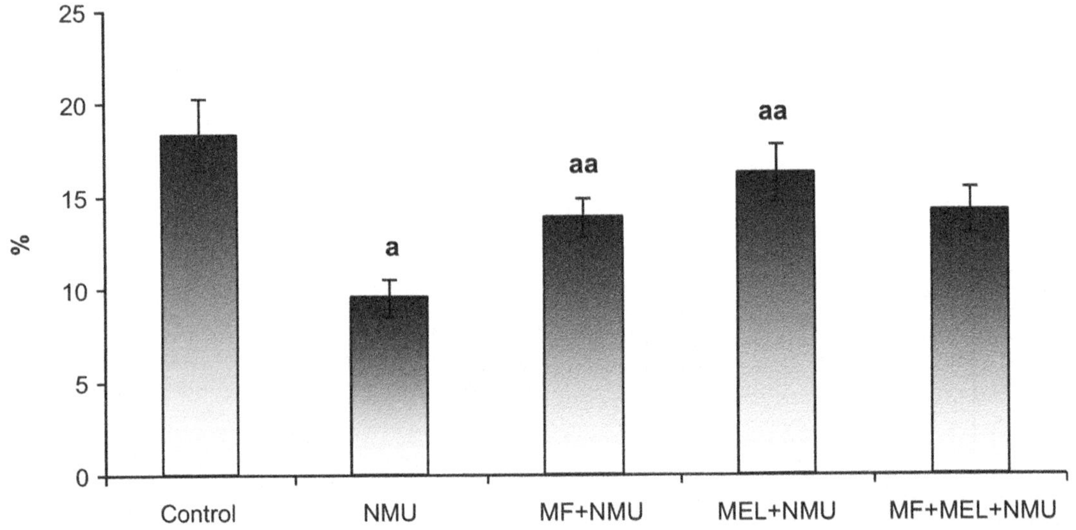

Fig. 3 Percentage of TAS changes in the heart tissue of female Sprague-Dawley rats subjected to a high fat diet in the mammary carcinogenesis model induced by N-methyl-N-nitrosourea (NMU). Metformin (MF) or melatonin (MEL) was administered alone and in combination (MF + MEL). Values are means ±SE. **a** – NMU group *vs.* control group; **aa** – MF + NMU and MEL + NMU groups *vs.* NMU group; **bb** – MF + MEL + NMU group *vs.* MF + NMU and MEL + NMU group. Significant differences between groups were designated as follows: **a** – NMU group *vs.* Control group; **aa** – MF + NMU and MEL + NMU group *vs.* NMU group

2003). In breast cancer, dysregulated ROS metabolism has been observed, detected by various indicators in plasma or blood cells, including red blood cells and platelets (Sosa et al. 2013; Kedzierska et al. 2012).

The present study demonstrates that NMU-induced mammary carcinogenesis featured an enhanced oxidative stress in the heart tissue of female Sprague-Dawley rats, as estimated from increased TBARS and oxidatively modified proteins. Oxidative stress has been proposed as a key mechanism in the toxic effects of carcinogenesis in many organs of the body, including the liver, heart, and kidney (Sekkin et al. 2015). The role of two main groups of oral antidiabetics consisting of biguanides and thiazolidinediones is debated in the literature. However, due to a risk of inducing lactic acidosis, only metformin has been used in clinical practice. The oncostatic effects of metformin

Table 2 Antioxidant enzyme activity in the heart tissue of female Sprague-Dawley rats subjected to the mammary carcinogenesis model induced by N-methyl-N-nitrosourea (NMU)

	Control	NMU	MF + NMU	MEL + NMU	MF + MEL+ NMU
SOD $U \cdot mg^{-1}$ protein	322.87 ± 41.76	487.22 ± 22.98[a]	571.21 ± 82.42	354.33 ± 31.22[aa]	675.22 ± 36.05 [bb]
CAT $\mu mol \cdot min^{-1} \cdot mg^{-1}$ protein	22.16 ± 1.89	12.44 ± 1.54[a]	18.35 ± 2.44[aa]	16.76 ± 0.81[aa]	16.25 ± 3.44
GR nmol $NADPH_2 \cdot min^{-1} \cdot mg^{-1}$ protein	76.33 ± 5.23	98.56 ± 9.27[a]	124.89 ± 9.87[aa]	88.13 ± 6.25	90.33 ± 9.14
GPx nmol $GSH \cdot min^{-1} \cdot mg^{-1}$ protein	82.14 ± 5.85	66.47 ± 5.84[a]	147.12 ± 8.47[aa]	160.85 ± 22.14 [aa]	111.82 ± 18.24 [bb]

Metformin (MF) or melatonin (MEL) was administered alone and in combination (MF + MEL)
Values are means ±SE. Significant differences between groups are designated as follows: [a]NMU group *vs.* Control group; [aa]MF + NMU and MEL+ NMU group *vs.* NMU group

are reported in numerous neoplasms (Zhu et al. 2015; Anisimov 2014), including a mammary carcinogenic model *in vivo* (Bojková et al. 2009).

In the present study, metformin with NMU exposure and a high fat diet reduced the oxidative stress in the heart tissue of rats, which was accompanied by a decrease in diene conjugates, and an elimination of ROS products as assessed from the TAS level. These results suggest that metformin reversed the effects of NMU on the oxidative stress, notably activating the CAT and GR proteins. Metformin exerts its effects through AMP-activated protein kinase activation and a subsequent mammalian target of rapamycin (mTOR) pathway inhibition. Metformin decreases the protein synthesis and cell proliferation. AMPK activation is also reported in the effects of thiazolidinediones, yet another group of antidiabetics that improve insulin resistance in diabetic patients. Thiazolidinediones are synthetic ligands of peroxisome proliferator-activated receptors. These receptors act as transcription factors and are involved in the immune response modulation, cell proliferation, and lipid transport and accumulation (He et al. 2012). These drugs may inhibit carcinogenesis through the cell cycle arrest, induction of apoptosis, suppression of angiogenesis, and anti-inflammatory activity. They also inhibit malignant growth, including mammary cancer cells, in numerous *in vitro* and *in vivo* models. We have previously found a prominent oncostatic effect of pioglitazone in chemically-induced mammary carcinogenesis (Bojková et al. 2010); rosiglitazone exerted only a partial effect in a similar experimental model (Bojková et al. 2014).

The present findings are in line with those of Favero et al. (2015), who have shown that melatonin reduces body weight, adipose tissue depots, adipocyte hyperplasia and hypertrophy, blood glucose, pro-inflammatory factors, and restores the adipokine physiological profile. We found that melatonin suppresses the pathophysiological mechanisms underlying the influence of a high fat diet and toxic effects of NMU. Several studies support the hypothesis that melatonin, a potent antioxidant, may be of benefit in a variety of conditions, acting as an anti-inflammatory, immunomodulatory, anti-proliferative, pro-apoptotic, and anti-angiogenic compound (Carrasco et al. 2015; Esrefoglu et al. 2012; Carpentieri et al. 2012; Jou et al. 2007; Reiter et al. 2003). These properties may underlie melatonin's oncostatic effect, first reported in the MCF-7 mammary adenocarcinoma cell line. The therapeutic effect of melatonin has been reported in many experimental neoplasms and in human cancers (Borin et al. 2016). The multifaceted effects of melatonin are mediated by increased activity of genes encoding the antioxidant SOD, CAT, GR, and GPx proteins (Okatani et al. 2000). Several authors have demonstrated the preventive effects of melatonin, administered alone or in combination with other agents, in female rat mammary carcinogenesis (Kubatka et al. 2014; Orendáš et al. 2014; Orendáš et al. 2009). The important finding of the present work is that such effects were greater than those exerted by metformin in the heart tissue in the NMU model of carcinogenesis.

Experimental findings support the oncostatic role of melatonin in the hormone-dependent mammary tumors (Borin et al. 2016). The mechanisms of the anti-tumor action of melatonin include interactions with the tumor cell estrogen-dependent pathways, cell cycle regulation, inhibited telomerase activity, modulation of fatty acid transport, and the metabolic and anti-invasive characteristics of a hormone (Viswanathan and Schernhammer 2009). Direct effects of melatonin, associated with the estrogen-dependent pathways of cellular metabolism, have been intensively studied *in vitro*, mainly in the cultured human breast tumor cells (MCF-7). These cells express receptors for melatonin, estrogen, and progesterone, and their growth is dependent on estrogen (Cos and Sanchez-Barcelo 2000). Estrogen-induced transcription and proliferation in the synchronized MCF-7 cells are inhibited by melatonin (Rato et al. 1999). Plausibly, cAMP and calmodulin constitute a link between the melatonin and estrogen signaling pathways. cAMP and other protein kinase activators stimulate estrogen-mediated transcription, through a mechanism involving the phosphorylation of estrogen receptors. In MCF-7 cells, estrogens activate adenylate cyclase *via* a non-transcriptional pathway and significantly increase the concentration of intracellular cAMP (Zivadinovic et al. 2005). In contrast, melatonin, acting through membrane receptors, inhibits adenylate cyclase and reduces the level of cAMP (Kiefer et al. 2002).

This study has some limitations related to the experimental methodology. Due to a high number of analyses, we resigned from the investigation of the effects exerted by a high fat diet alone. We believe that such evaluations are sufficiently present in the literature. Also, we believe we eliminated this factor by giving a high fat diet to all the rats investigated. This approach enabled us to obtain the data on the influence of metformin and melatonin on oxidative stress in the heart tissue of female Sprague-Dawley rats remaining on a high fat diet and subjected to the mammary carcinogenic process induced by NMU; the data that have not been hitherto reported. The evidence provided demonstrates that metformin or melatonin alone, and both combined, modulate various metabolic pathways of oxidative stress in the heart tissue. The results suggest that metformin is less effective than melatonin in reversing the effects of NMU and a high fat diet on oxidative stress. Thus, melatonin might be useful as supportive therapy in the conditions associated with elevated oxidative stress caused by a high fat diet and related carcinogenesis.

5 Conclusions

Mammary carcinogenesis induced by N-methyl-N-nitrosourea (NMU) and a high fat diet leads to increased oxidative stress by generating of ROS in the heart tissue of female Sprague-Dawley rats, as estimated from the enhanced content of TBARS, OMP (aldehyde and ketone derivatives), and antioxidant defense enzymes such as SOD, GR, and GPx. Metformin reduced the oxidative stress, which was accompanied by a decrease in diene conjugates and the elimination of ROS as indicated by a stable TAS level and the activation of CAT and GR enzymes. Administration of melatonin to the NMU-exposed rats staying on a high fat diet caused an increase in TAS through a substantial reduction in ROS generation, as estimated by OMP AD and OMP KD levels, and LPO processes in the initial (diene conjugates) and terminal stages (TBARS) and by increases in the antioxidant enzymes CAT and GPx. A combination of metformin and melatonin reversed the effect of NMU toxicity on oxidative stress. Finally, melatonin acted significantly stronger than metformin in oxidative stress reduction in the heart tissue, caused by NMU carcinogenesis and a high fat diet.

Acknowledgements We thank Ingrid Obšitošova, Eva Petrovičova, Tomaš Raši, and Monika Kassayova for excellent technical assistance. This work was supported by the Science Grant Agency VEGA, No.1/0153/13 of the Ministry of Education in Slovakia. The study was in part supported by the Pomeranian University of Slupsk, the National State University of Chernihiv, the Medical University of Gdansk, the Medical University of Warsaw, and Polish National Commission for UNESCO.

Conflicts of Interest The authors declare no conflicts of interest related to this article.

References

Anisimov VN (2014) Do metformin a real anticarcinogen? A critical reappraisal of experimental data. Ann Trans Med 2:60–65

Bojková B, Orendáš P, Garajová M, Kassayová M, Kútna V, Ahlersová E, Ahlers I (2009) Metformin in chemically-induced mammary carcinogenesis in rats. Neoplasma 56:267–272

Bojková B, Garajová M, Kajo K, Péč M, Kubatka P, Kasasyová M, Kisková T, Orendáš P, Ahlersová E, Ahlers I (2010) Pioglitazone in chemically induced mammary carcinogenesis in rats. Eur J Cancer Prev 19:379–384

Bojková B, Orendáš P, Kubatka P, Péč M, Kassayová M, Kisková T, Kajo K (2014) Positive and negative effects of glitazones in carcinogenesis: experimental models *vs.* clinical practice. Pathol Res Pract 210:465–472

Borin TF, Arbab AS, Gelaleti GB, Ferreira LC, Moschetta MG, Jardim-Perassi BV, Iskander A, Varma NR, Shankar A, Coimbra VB, Fabri VA, de Oliveira JG, de Campos Zuccari DA (2016) Melatonin decreases breast cancer metastasis by modulating ROCK-1 expression. J Pineal Res 60:3–15

Bradford MM (1976) A rapid and sensitive method for the quantitation of microgram quantities of protein utilizing the principle of protein-dye binding. Anal Biochem 72:248–254

Carpentieri A, de Barboza GD, Areco V, Lopez MP, de Talamoni NT (2012) New perspectives in melatonin uses. Pharmacol Res 65:437–444

Carrasco C, Rodrigues AB, Pariente JA (2015) Melatonin as a stabilizer of mitochondrial function: role in diseases and aging. Turk J Biol 39:822–831

Cos S, Sanchez-Barcelo EJ (2000) Melatonin, experimental basis for a possible application in breast cancer prevention and treatment. Histol Histopathol 15:637–647

Didkowska J, Wojciechowska U, Didkowska J, Zatonski W (2007) Cancer in Poland in 2005. The Maria Sklodowska-Curie Memorial Cancer Centre, Department of Epidemiology and Cancer Prevention, Warsaw, pp. 86–87

Dubinina EE, Burmistrov SO, Khodov DA, Porotov IG (1995) Oxidative modification of human serum proteins. A method of determining it. Vopr Med Khim 41:24–26

Erdem SB, Nacaroglu HT, Bag O, Karkıner CSU, Korkmaz HA, Can D (2015) DRESS syndrome associated with type 2 diabetes in a child. Cent Eur J Immunol 40:493–496

Esrefoglu M, Iraz M, Ates B, Gul M (2012) Melatonin and CAPE are able to prevent the liver from oxidative damage in rats: an ultrastructural and biochemical study. Ultrastruct Pathol 36:171–178

Favero G, Stacchiotti A, Castrezzati S, Bonomini F, Albanese M, Rezzani R, Rodella LF (2015) Melatonin reduces obesity and restores adipokine patterns and metabolism in obese (ob/ob) mice. Nutr Res 35:891–900

Ferlay J, Steliarova-Foucher E, Lortet-Tieulent J, Rosso S, Coebergh JW, Comber H, Forman D, Bray F (2013) Cancer incidence and mortality patterns in Europe: estimates for 40 countries in 2012. Eur J Cancer 49:1374–1403

Galaktionova LP, Molchanov AV, Elchaninova SA, Varshavskiy BY (1998) Lipid peroxidation in patients with gastric and duodenal ulcers. Klinicheskaia Labaratornaia Diagnostika 6:10–14

Glatzle D, Vuilleumier JP, Weber F, Decker K (1974) Glutathione reductase test with whole blood, a convenient procedure for the assessment of the riboflavin status in human. Experientia 30:665–666

Goodwin PJ, Stambolic V (2011) Obesity and insulin resistance in breast cancer-chemoprevention strategies with a focus on metformin. Breast Suppl 3:S31–S355

Goodwin PJ, Thompson AM, Stambolic V (2012) Diabetes, metformin, and breast cancer: lilac time? J Clin Oncol 30:2812–2814

He X, Esteva FJ, Ensor J, Hortobagyi GN, Lee MH, Yeung SC (2012) Metformin and thiazolidinediones are associated with improved breast cancer-specific survival of diabetic women with HER2+ breast cancer. Ann Oncol 23:1771–1780

Jou MJ, Peng TI, PZ Y, Jou SB, Reiter RJ, Chen JY, HY W, Chen CC, Hsu LF (2007) Melatonin protects against common deletion of mitochondrial DNA augmented oxidative stress and apoptosis. J Pineal Res 43:389–403

Kamyshnikov VS (2004) Reference book on clinic and biochemical researches and laboratory diagnostics. MEDpress-inform, Moscow

Kedzierska M, Olas B, Wachowicz B, Jeziorski A, Piekarski J (2012) Relationship between thiol, tyrosine nitration and carbonyl formation as biomarkers of oxidative stress and changes of hemostatic function of plasma from breast cancer patients before surgery. Clin Biochem 45:231–236

Kiefer T, Ram PT, Yuan L, Hill SM (2002) Melatonin inhibits estrogen receptor transactivation and cAMP levels in breast cancer cells. Breast Cancer Res Treat 71:37–45

Koroliuk MA, Ivanova LI, Majorova IG, Tokarev VE (1988) A method of determining catalase activity. Lab Delo 1:16–19

Kostiuk VA, Potapovich AI, Kovaleva ZV (1990) A simple and sensitive method of determination of superoxide dismutase activity based on the reaction of quercetin oxidation. Vopr Med Khim 36:88–91

Kubatka P, Kalická K, Chamilová M, Ahlersová E, Ahlers I, Bojková B, Adámeková E (2002) Nimesulide and melatonin in mammary carcinogenesis prevention in female Sprague-Dawley rats. Neoplasma 49:255–259

Kubatka P, Bojková B, Kassayová M, Orendáš P, Kajo K, Výbohová D, Kružliak P, Adamicová K, Péč M, Stollárová N, Adamkov M (2014) Combination of pitavastatin and melatonin shows partial antineoplastic effects in a rat breast carcinoma model. Acta Histochem 116:1454–1461

Kurhaluk N, Sliuta A, Kyriienko S, Winklewski P (2017) Melatonin restores white blood cell count, diminishes glycated haemoglobin level and prevents liver, kidney and muscle oxidative stress in mice exposed to acute ethanol intoxication. Alcohol Alcohol 52:521–528

Leone A, Di Gennaro E, Bruzzese F, Avallone A, Budillon A (2014) New perspective for an old antidiabetic drug: metformin as anticancer agent. Cancer Treat Res 159:355–376

Levine RL, Garland D, Oliver CN, Amici A, Climent I, Lenz AG, Ahn BW, Shaltiel S, Stadtman ER (1990) Determination of carbonyl content in oxidatively modified proteins. Methods Enzymol 186:465–478

Moin VM (1986) A simple and specific method for determining glutathione peroxidase activity in erythrocytes. Lab Delo 12:724–727

Okatani Y, Wakatsuki A, Kaneda C (2000) Melatonin increases activities of glutathione peroxidase and superoxide dismutase in fetal rat brain. J Pineal Res 28:89–96

Orendáš P, Kassayova M, Kajo K, Ahlers I, Kubatka P, Bojkova B, Pec M, Ahlersova E (2009) Celecoxib and melatonin in prevention of female rat mammary carcinogenesis. Neoplasma 56:252–258

Orendáš P, Kubatka P, Bojková B, Kassayová M, Kajo K, Výbohová D, Kružliak P, Péč M, Adamkov M, Kapinová A, Adamicová K, Sadloňová V, Chmelová M, Stollárová N (2014) Melatonin potentiates the anti-tumour effect of pravastatin in rat mammary gland carcinoma model. Int J Exp Pathol 95:401–410

Panis C, Victorino VJ, Herrera AC, Freitas LF, De Rossi T, Campos FC (2012) Differential oxidative status and immune characterization of the early and advanced stages of human breast cancer. Breast Cancer Res Treat 133:881–888

Pollak M (2013) Potential applications for biguanides in oncology. J Clin Invest 123:3693–3700

Popko K, Gorska E, Demkow U (2010) Influence of interleukin-6 and G174C polymorphism in IL-6 gene on obesity and energy balance. Eur J Med Res 15 (Suppl 2):123–127

Rato AG, Pedrero JG, Martinez MA, del Rio B, Lazo PS, Ramos S (1999) Melatonin blocks the activation of estrogen receptor for DNA binding. FASEB J 13:857–868

Reiter RJ, Tan DX, Mayo JC, Sainz RM, Leon J, Czarnocki Z (2003) Melatonin as an antioxidant: biochemical mechanisms and pathophysiological implications in humans. Acta Biochim Pol 50:1129–1146

Ren DL, Sun AA, Li YJ, Chen M, Ge SC, Hu B (2015) Exogenous melatonin inhibits neutrophil migration through suppression of ERK activation. J Endocrinol 227:49–60

Sekkin S, Duygu Ipek E, Boyacioğlu M, Kum C, Karademir U, Sultan Yalinkilinç H, Onur Ak M, Başaloğlu H (2015) DNA protective and antioxidative effects of melatonin in streptozotocin-induced diabetic rats. Turkish J Biol 39:2322–2340

Shin IS, Shin NR, Park JW, Jeon CM, Hong JM, Kwon OK, Kim JS, Lee IC, Kim JC, SR O, Ahn KS (2015) Melatonin attenuates neutrophil inflammation and mucus secretion in cigarette smoke-induced chronic obstructive pulmonary diseases via the suppression of Erk – Sp1 signaling. J Pineal Res 58:50–60

Sosa V, Moline T, Somoza R, Paciucci R, Kondoh H, Me LL (2013) Oxidative stress and cancer: an overview. Ageing Res Rev 12:376–390

Sudnikovich EJ, Maksimchik YZ, Zabrodskaya SV, Kubyshin VL, Lapshina EA, Bryszewska M, Reiter RJ, Zavodnik IB (2007) Melatonin attenuates metabolic disorders due to streptozotocin-induced diabetes in rats. Eur J Pharmacol 569:180–187

Syczewska-Weber K, Rucinski P (2008) The main challenges of polish oncology. Public Heath Rep 123:655–663

Tas F, Hansel H, Belce A, Ilvan S, Argon A, Camlica H (2005) Oxidative stress in breast cancer. Med Oncol 22:11–15

Viswanathan AN, Schernhammer ES (2009) Circulating melatonin and the risk of breast and endometrial cancer in women. Cancer Lett 281:1–7

Winklewski PJ, Radkowski M, Wszedybyl-Winklewska-M, Demkow U (2015) Brain inflammation and hypertension: the chicken or the egg? J Neuroinflammation 12:85

Zhu Z, Jiang W, Thompson MD, Echeverria D, McGinley JN, Thompson HJ (2015) Effects of metformin, buformin, and phenformin on the post-initiation stage of chemically induced mammary carcinogenesis in the rat. Cancer Prev Res 8:518–527

Zivadinovic D, Gametchu B, Watson CS (2005) Membrane estrogen receptor-alpha levels in MCF-7 breast cancer cells predict cAMP and proliferation responses. Breast Cancer Res 7:101–112

Advs Exp. Medicine, Biology - Neuroscience and Respiration (2018) 38: 21–30
DOI 10.1007/5584_2017_122

Published online: 19 November 2017

Evaluation of the Implementation of Hospital Antibiotic Policy in Oncosurgery Ward: A Six-Year Experience

Aneta Nitsch-Osuch, Damian Okruciński, Magdalena Dawgiałło, Izabela Gołębiak, and Ernest Kuchar

Abstract

The hospital antibiotic policy should be implemented to rationalize the antibiotic use and to decrease the risk of spread of resistant bacteria. The aim of this study was to describe the antibiotic consumption patterns in a single oncosurgery ward before and after the implementation of hospital antibiotic policy. We conducted a retrospective analysis of the antibiotic use at the oncosurgery ward in Warsaw, Poland, in the years 2011–2016. Calculations were based on daily defined doses (DDD), DDD/100 hospitalizations, and DDD/100 person-days. Drug utilization rates (DU 90% and DU 100%) were also analyzed. After the implementation of hospital antibiotic policy, a total antibiotic consumption increased (365.35 DDD in 2011 *vs.* 1359.22 DDD in 2016). A significant change was observed in the antibiotic consumption patterns: the use of amoxicillin clavulanate and carbapenems or glycopeptides decreased significantly ($p < 0.05$), while the use of ciprofloxacin and aminoglycosides increased ($p < 0.05$). The DU100% rate varied from 6 in 2011 to 12 in 2016; while DU 90% rate varied from 2 in 2011 to 3–5 in 2013–2016. Although the implementation of hospital antibiotic policy did not result in a decrease in the antibiotic consumption, it seems to provide a favorable change into the antibiotic consumption pattern.

Keywords

Antibiotic resistance · Antibiotic stewardship · Antibiotic use pattern · Hospital antibiotic policy · Oncosurgery ward

1 Introduction

Bacterial resistance to antibiotics is best illustrated by the occurrence of the methicillin-

A. Nitsch-Osuch (✉)
Department of Social Medicine and Public Health, Medical University of Warsaw, Warsaw, Poland

St. Family Hospital, Warsaw, Poland
e-mail: anitsch@wum.edu.pl

D. Okruciński and E. Kuchar
Department of Pediatrics and Clinical Assessment Unit, Medical University of Warsaw, Warsaw, Poland

M. Dawgiałło and I. Gołębiak
Department of Social Medicine and Public Health, Medical University of Warsaw, Warsaw, Poland

resistant *Staphylococcus aureus* (MRSA), vancomycin-resistant *Enterococcus*, extended-spectrum beta-lactamases (ESBL), and carbapenemase-producing bacteria *Klebsiella pneumoniae* (Davies and Davies 2010; Schwaber and Carmelli 2008). In light of the growing drug resistance, the development of new antibacterial substances is not a promising occurrence. Moreover, breakthroughs in the area of research in the next few years are rather improbable. Therefore, a rational antibiotic therapy aiming at decreasing risk of drug resistance should be implemented on a large scale, in both hospitals and outpatient settings. An example of this activity might be the implementation of the hospital antibiotic policy. The aim of this policy is to introduce a rational antibiotic therapy and do it with the utmost consideration for patient care. Antibiotics should be prescribed in proper doses, for optimum duration of therapy, with a minimum risk of adverse effects or the development of antibiotic resistance, and with a consideration for the lowest possible therapy cost (Dik et al. 2015; Davey et al. 2005). The aim of this study was to retrospectively evaluate the implementation of hospital antibiotic policy in the Oncosurgery Ward during a 6-year observation time, a subject that was hitherto only sparingly dealt with in the relevant medical literature.

2 Methods

The study was approved by a local Bioethics Committee of Warsaw Medical University in Warsaw, Poland. A retrospective quantitative and qualitative analysis of the antibiotic use, before and in the consecutive 5 years after the implementation of the hospital antibiotic policy, was conducted in the Oncosurgery Ward of the St. Family Hospital in Warsaw. The hospital is a second level referral hospital, and the Oncosurgery Ward has 20 beds. The number of hospitalizations and patient-days in the analyzed years are presented in Table 1. The ward specializes in breast cancer surgery and thyroid surgeries. Surgical procedures performed in the period above outlined are summarized in Table 2. There was a shift towards breast cancer surgery, whereas thyroid surgery declined.

Responsibility for the implementation of the hospital antibiotic policy was placed on the Infection Control Team that consisted of an epidemiologist, pharmacist, microbiologist, and epidemiological nurse. One of the key goals of the policy was to create a Hospital Antibiotics List divided into three groups: 'first-line' antibiotics (prescribed by all physicians at all times, e.g. cefazolin, amoxicillin, and doxycycline); 'controlled antibiotics' (prescribed by all physicians, but a continuation of therapy required a permission from the head physician,

Table 1 Patients traffic in the Oncosurgery Ward before (2011) and after (2012–2016) the implementation of the hospital antibiotic policy

Year	2011	2012	2013	2014	2015	2016
Number of hospitalized patients	1408	2101	1078	1153	1172	1156
Number of patient-days	2558	3171	2972	4411	3820	3749
Average hospital stay (days)	1.82	1.50	2.73	3.84	3.24	3.24

Table 2 Proportion of surgical procedures in the Oncosurgery Ward before (2011) and after (2012–2016) the implementation of the hospital antibiotic policy

Year	2011	2012	2013	2014	2015	2016
Breast surgery	31%	51%	66%	58%	68%	72%
Gastrointestinal surgery	2%	4%	4%	3%	5%	2%
Thyroid surgery	27%	9%	3%	7%	5%	7%
Others	40%	36%	27%	32%	22%	19%

e.g., amoxicillin clavulanate, cefuroxime, ceftriaxone, clindamycin, ciprofloxacin, piperacillin with tazobactam, gentamycin, amikacin); and 'restricted antibiotics', prescribed with a written permission of the Infection Control Team head, e.g., carbapenems (imipenem and meropenem) and glycopeptides (teicoplanin and vancomycin).

We evaluated the antibiotic use according to the methodology described previously (Nitsch-Osuch et al. 2015a, b). The evaluation included the Hospital Pharmacy data consisting of the ordered and prescribed antibiotics, the type of antibiotic, the route of administration, and the dose as well as the Medical Statistics Department data consisting of the number of hospitalizations and patient-days. Moreover, antibiotic consumption pattern was defined, specifying the proportion of antibiotics used in terms of daily defined dose (DDD), which is the average daily dose of a drug, specified with a code of the Anatomical Therapeutic Chemical Classification, used in therapy as the main indication. The following formula was employed: DDD = (number of doses × g/dose) / DDD value (g). Other parameters used in the assessment of antibiotic use were as follows:

- DDD/100 hospitalizations;
- DDD/100 person-days;
- Drug Utilization 100% (DU 100%);
- Drug Utilization 90% (DU 90%).

The DDD/100 hospitalizations and DDD/100 person-days enable the inter-departmental comparisons and also describe the trends in antibiotic consumption in a ward at different time intervals. The DU100% and DU90% are considered to reflect a correctness of antibiotic treatment application. The DU100% shows how many groups of antibiotics are used, while DU90% shows which groups of antibiotics comprise 90% of the antibiotics used.

Statistical evaluation was performed using a chi-squared test. A p-value of <0.05 defined statistically significant differences. We also calculated the odds ratios (OR) and 95% confidence intervals (95%CI) using an online statistical calculator available at www.medcalc.3000.

3 Results

Consumption of antibiotics was steadily increasing; from 365.4 DDD in 2011 before the implementation of the hospital antibiotic policy to 1359.2 DDD in the fifth year of observation on the hospital antibiotic policy in 2016 (Table 3). A similar upward trend concerned also the use of antibiotics expressed in DDD/100 person-days (Table 4) and DDD/100 hospital admissions (Table 5). There was a significant change in the antibiotic consumption pattern after the implementation of the hospital antibiotic policy in 2012–2016 (Fig. 1). There was a decrease in the use of penicillin with inhibitors, mainly amoxicillin clavulanate, 64% DDD in 2011 *vs.* 26.5% in 2016, and cephalosporins (mainly cefuroxime and ceftriaxone), 22.5% DDD in 2011 *vs.* 6.6% in 2016. Furthermore, the use of restricted antibiotics (carbapenems and glycopeptides) decreased from 7.0% in 2011 *vs.* 1.0% in 2016. The most spectacular change was the increased consumption of ciprofloxacin from 5.0% in 2011 to 57.3% in 2016 and aminoglycosides from 0% in 2011 to 8.5% in 2016. The detailed antibiotic consumption patterns period are presented in Table 6. The DU 100% rate rose initially after the implementation of the hospital antibiotic policy HAP; in 2011 it amounted to 6 antibiotics, whereas in 2014 it was 11 antibiotics, only to drastically decrease in 2015 (DU100% 5). The DU ratio of 90% were 2 in 2011, and then 3 or 5 antibiotics in 2012–2016 (Table 7).

Table 3 .Antibiotic use before (2011) and after (2012–2016) the implementation of the hospital antibiotic policy in terms of the total DDD consumed

Year	2011	2012	2013	2014	2015	2016
Antibiotic	DDD					
Penicillins	0	0	29	0	0	2
Amoxicillin	0	0	29	0	0	0
Ampicillin	0	0	0	0	0	2
Penicillins with inhibitors					350,8	366.2
Amoxicillin clavulanate	234.6	464.2	380.8	341.6	350.8	366.2
Piperacillins with tazobactam	0	0	0	2.1	0	6.1
Cephalosporins	80.8	146.6	116.3	209.0	60.6	89.8
I Generation Cephazolin	0	0	0	32.7	22	2
II Generation Cefuroxime	78.75	116.6	101.3	131.3	13.6	72.8
III Generation Ceftriaxone	2	30	15	45	25	15
Macrolides	0	0	0	0	0	5
Azythromycin	0	0	0	0	0	5
Lincosamides	5	10.6	19	94.5	0	3.8
Clindamycin	5	10.6	19	94.5	0	3.8
Fluoroquinolones	20	5	90	640	770	780
Ciprofloxacin	20	5	90	640	770	780
Tetracyclines	0	0	0	20	0	0
Doxycycline	0	0	0	20	0	0
Aminoglycosides	0	36.6	197.3	93.0	86.5	98.5
Gentamicin	0	36.6	133.3	0	0	0
Amikacin	0	0	64	93	86.5	98.5
Carbapenems	25	17.5	0	7.5	0	7.5
Imipenem	25	0	0	7.5	0	0
Meropenem	0	17.5	0	0	0	7.5
Glycopeptides	0	0	0	3.8	0	6.5
Vancomycin	0	0	0	3.8	0	6.5
Summary	365.4	680.5	843.9	1411.4	1267.9	1359.2

4 Discussion

4.1 Effects of Implementation of the Hospital Antibiotic Policy on Antibiotic Consumption Patterns

The implementation of the hospital antibiotic policy did not reduce a total consumption of antibiotics in the ward. On the contrary, antibiotic consumption steadily increased. This seemingly surprising observation may be caused by a different profile of patients being admitted to the ward. While before the HAP implementation, mainly thyroidectomies and mastectomies were carried out, breast cancer surgeries predominated after the implementation. Patients after mastectomy often required antibiotic prophylaxis against infections, notably with ciprofloxacin, or antibiotic treatment for nosocomial, mainly surgical site, and infections. According to HAP, ciprofloxacin was prescribed for 5–7 days after surgery, especially for patients who had expanders. Increased antibiotic use could also be linked to a longer hospital stay in the time after HAP implementation in subsequent years (Table 1). The results observed were not influenced by the more frequent use of cephazoline as a preoperative antibiotic, since this antibiotic was given in the surgical theater room, not in the ward.

Antibiotic consumption in our ward was lower than that in similar oncological hospitals in France (39 DDD/100 person-days) (Dumarin

Table 4 Antibiotic use before (2011) and after (2012–2016) the implementation of the hospital antibiotic policy in terms of the number of consumed DDD/100 person-days

Year	2011	2012	2013	2014	2015	2016
Antibiotic	DDD/100 person-days					
Penicillins	9.2	14.6	13.8	7.8	0	0.1
Amoxicillin	0	0	1.0	7.7	0	0
Ampicillin	0	0	0	0	0	0.1
Penicillins with inhibitors					9.1	96.2
Amoxicillin clavulanate	9.2	14.6	12.8	0.1	9.1	96.0
Piperacillins with tazobactam	0	0	0	0.2	0	0.2
Cephalosporins	3.1	4.6	3.9	4.7	1.5	535.9
I Generation Cephazolin	0	0	0	0.7	0.5	0.1
II Generation Cefuroxime	3	3.7	3.4	3.0	0.4	535.4
III Generation Ceftriaxone	0.1	0.9	0.5	1.0	0.7	0.4
Macrolides	0	0	0	0	0	0.1
Azythromycin	0	0	0	0	0	0.1
Lincosamides	0.2	3.3	0.6	2.1	0	0.1
Clindamycin	0.2	3.3	0.6	2.1	0	0.1
Fluoroquinolones	0.8	0.2	3	15.0	20.2	20.8
Ciprofloxacin	0.8	0.2	3	15.0	20.2	20.8
Aminoglycosides	0	1.2	6.6	2.1	2.3	2.6
Gentamicin	0	1.2	4.5	0	0	0
Amikacin	0	0	2.1	2.1	2.3	2.6
Carbapenems	1.0	0.6	0	0.2	0	0.2
Imipenem	1.0	0	0	0.2	0	0
Meropenem	0	0.6	0	0	0	0.2
Glycopeptides	0	0	0	0.1	0	0.2
Vancomycin	0	0	0	0.1	0	0.2
Summary	14.2	24.4	27.8	31.5	33.2	36.3

et al. 2010), which can be explained by the predominance of planned admissions in our ward. Antibiotic consumption is greater in wards having more than 25% of 'acute admission' beds (Miliani et al. 2008). Likewise, in an Iranian study of Ayazi et al. (2015) antibiotic consumption has been greater than that in the present study; it amounted to 5–7 DDD/100 person-days and showed an upward trend in the follow-up period. In line with the present findings, HAP implementation did not result in a lower antibiotic consumption pattern in a Chinese study of Bao et al. (2015). Previous Polish studies related to other than surgical profiles have also provided data indicating a growing trend in the total DDD use of antibiotics after HAP implementation (Nitsch-Osuch et al. 2015a, b).

In the present study, HAP implementation had a significant effect on the scope of antibiotics used: consumption of penicillin with inhibitors, mainly amoxicillin clavulanate, of cephalosporins, mainly cefuroxime, and of restricted antibiotics, mainly carbapenems and glycopeptides, decreased, whereas the use of ciprofloxacin and aminoglycosides increased. All these changes were statistically significant and in our opinion they were in a positive direction in terms of antibiotic consumption patterns. Firstly, a reduction in the use of carbapenems may be viewed as a favorable result of HAP. Carbapenems are wide spectrum antibiotics often called 'last chance treatment'. Zilberberg et al. (2017) have shown that the infection with carbapenemase resistant *Enterobacteriaceae* is associated with a fourfold increased risk of receiving previous inappropriate empiric treatment, which, in turn, increases mortality, length of hospitalization, and costs. Increased

Table 5 Antibiotic use before (2011) and after (2012–2016) the implementation of the hospital antibiotic policy in terms of the total number of consumed DDD/100 hospitalizations

Year	2011	2012	2013	2014	2015	2016
Antibiotic	DDD/100 hospitalizations					
Penicillins	0	0	0.4	0	0	0.2
Amoxicillin	0	0	0.4	0	0	0
Ampicillin	0	0	0	0	0	0.2
Penicillins with inhibitors					29.9	36.1
Amoxicillin clavulanate	16.6	30.75	35	29.6	29.9	31.1
Piperacillins with tazobactam	0	0	0	0.2	0	0.5
Cephalosporins	5.7	7.9	10.7	18.1	5.2	7.8
I Generation Cephazolin	0	0	0	2.8	1.9	0.2
II Generation Cefuroxime	5.6	5.5	9.3	11.4	1.2	6.3
III Generation Ceftriaxone	0.1	2.4	1.4	3.9	2.1	1.3
Macrolides	0	0	0	0	0	0.4
Azythromycin	0	0	0	0	0	0.4
Lincosamides	0.4	0.5	1.7	8.2	0	0.3
Clindamycin	0.4	0.5	1.7	8.2	0	0.3
Fluoroquinolones	1.4	0.2	8.3	55.5	65.6	67.5
Ciprofloxacin	1.4	0.2	8.3	55.5	65.6	67.5
Tetracyclines	0	0	0	1.7	0.2	0
Doxycycline	0	0	0	1.7	0.2	0
Aminoglycosides	0	1.7	18.1	8	7.3	0.8
Gentamicin	0	1.7	12.2	0	0	0
Amikacin	0	0	5.9	8	7.3	0.8
Carbapenems	1.8	0.8	0	0.7	0	0.7
Imipenem	1.8	0	0	0.7	0	0
Meropenem	0	0.8	0	0	0	0.7
Glycopeptides	0	0	0	0.3	0	0.6
Vancomycin	0	0	0	0.3	0	0.6
Summary	25.9	41.9	74.2	122.2	108.2	117.6

consumption of carbapenems is associated with increased rate of carbapenem-resistant bacteria, including *Acinetobacter baumannii,* as observed by Mascarello et al. (2017). Secondly, a lower use of betalactam antibiotics, including cephalosporins of the second and third generations and penicillin with inhibitors could be viewed as a favorable result as well. The excessive use of third generation cephalosporins is a risk factor for the occurrence and spread of extended-spectrum beta-lactamases (ESBL)-producing bacterial strains. Effective control of their use is associated with a successful restriction in the spread of resistant bacteria (Kim et al. 2008; Urbánek et al. 2007) and of *Clostridium difficile* infections (Sullivan 2014).

Consumption of ciprofloxacin in the ward was higher as compared to hospitals of a similar profile in Great Britain. Schelenz et al. (2015) have shown that ciprofloxacin consumption is 14.9/100 hospitalizations, whereas in our ward it was 65.6 DDD/100 hospitalizations. Ciprofloxacin is a widely used antibiotic in hematology and oncology wards, but its overuse is conducive to antibiotic resistance among Gram-negative bacteria. Hence, it is advisable to rotate its use (Schelenz et al. 2015). In the light of these results, it appears appropriate to conclude that

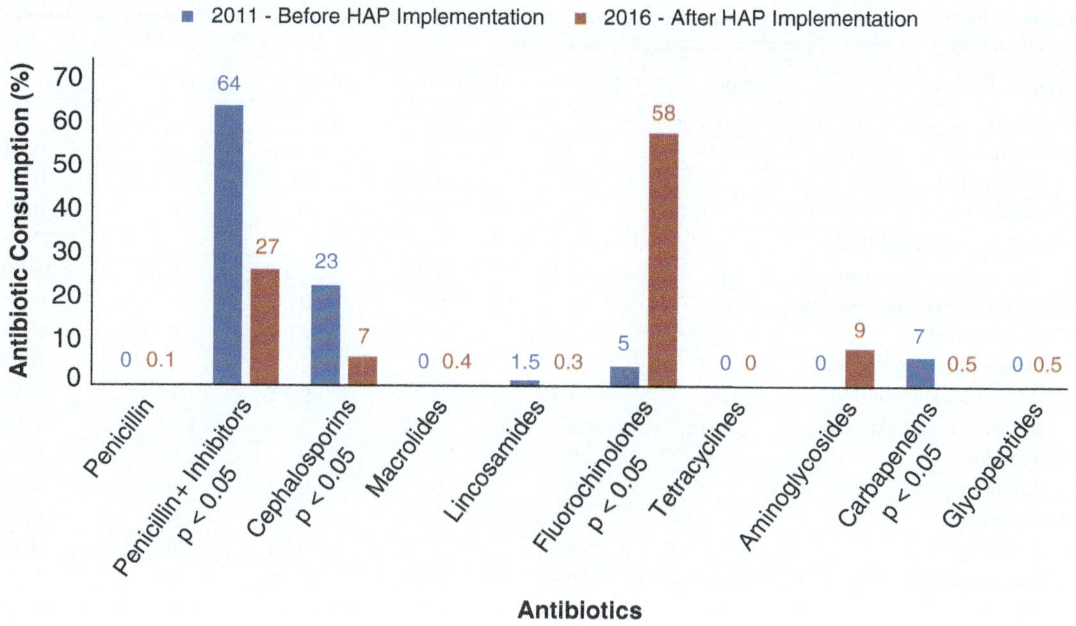

Fig. 1 Proportion of antibiotic consumption before (2011) and after (2012–2016) the implementation of the hospital antibiotic policy (HAP)

in our ward, ciprofloxacin may be an overused antibiotic, and this hypothesis is backed by the literature data. Hammuda et al. (2013) have shown that fluoroquinolones constitute 9% of the antibiotics consumed in a ward of similar profile, which is significantly less than the 65.5% noted in 2015 in the present study. A wide use of ciprofloxacin in our ward may be explained by the procedure of using this antibiotic in all patients who underwent breast prosthesis or were being prepared for breast prosthesis with skin expanders. The loss of prosthesis due to hospital acquired infection is one of the most dramatic and serious complications of the breast reconstruction surgery. The efficacy of ciprofloxacin for infection treatment after breast augmentation is well known. It has been shown that 40% of bacteria isolated in patients with a surgical site infection is resistant to commonly used antibiotics, but 86% of strains are sensitive to gentamicin, and 63% strains were sensitive to ciprofloxacin (Weichman et al. 2013).

4.2 Effects of Implementation of the Hospital Antibiotic Policy on the DU90% and DU100% Indicators

A positive result of HAP implementation in our ward also was the increased DU90% index, from 2 antibiotics in 2011 to 3–5 antibiotics in the years 2013–2016. Changes in the profile of patients admitted to the ward afore-mentioned and the creation and implementation of recommendations regarding infection treatment could lead to a growth of this index. The use of only a few antibiotics in the ward may suggest a suboptimal compliance to the antibiotic therapy recommendations, which may result in antibiotic resistance build-up (Birger et al. 2015; Roca et al. 2015). In this context, the reduction in the DU90% index to three in 2015 should be considered alarming, as it implies the use of too narrow a range of antibiotics in the ward.

Table 6 Patterns of the antibiotic consumption before (2011) and after (2012–2016) the implementation of the hospital antibiotic policy. in terms of prescribed antibiotic proportions

Year	2011	2012	2013	2014	2015	2016
Antibiotic						
Penicillins	0	0	3.4%	24.4%	0	0.1%
Amoxicillin	0	0	3.4%	0	0	0
Ampicillin	0	0	0	0	0	0.1%
Penicillins with inhibitors						27%
Amoxicillin clavulanate	64%	68.2%	45.2%	24.2%	27.6%	26.5%
Piperacillins with tazobactam	0%	0	0	0.2%	0%	0.5%
Cephalosporins	22.5%	21.5%	23.2%	14.8%	4.7%	6.6%
I Generation Cephazolin	0	0	0	2.3%	1.7%	0.1%
II Generation Cefuroxime	22%	17.5%	12%	9.3%	1%	5.4%
III Generation Ceftriaxone	0.5%	4%	11.3%	3.2%	2%	1.1%
Macrolides	0	0	0	0	0	0.4%
Azythromycin	0	0	0	0	0	0.4%
Lincosamides	1.5%	1.6%	2.2%	6.7%	0	0.3%
Clindamycin	1.5%	1.6	2.2%	6.7%	0	0.3%
Fluoroquinolones	5%	0.7%	10.7%	45.4%	61%	57.3%
Ciprofloxacin	5%	0.7%	10.7%	45.4%	61%	57.3%
Tetracyclines	0	0	0	1.4%	0	0
Doxycycline	0	0	0	1.4%	0	0
Aminoglycosides	0	5.3%	23.4%	6.6%	6.7%	8.5%
Gentamicin	0	5.3%	15.8%	0	0%	8.5%
Amikacin	0	0%	7.7%	6.6%	6.7%	0
Carbapenems	7%	2.6%	0	0.5%	0	0.6%
Imipenem	7%	0%	0	0.5%	0	0
Meropenem	0	2.6%	0	0	0	0.6%
Glycopeptides	0	0	0	0.3%	0	0.5%
Vancomycin	0	0	0	0.3%	0	0.5%
Summary	100%	100%	100%	100%	100%	100%

4.3 Study Strength and Weakness

The strength of our study is that we obtained results that can be used to compare the antibiotic consumption patterns in wards of similar treatment profile in Poland as well as in other countries. We show that HAP implementation provided favorable changes in the pattern of antibiotic consumption, even when a total number of antibiotic doses did not decrease.

One of the main limitations of the study is the methodology used to determine the antibiotic consumption, i.e., the calculation based on defined daily doses (DDD). This method is recommended by the WHO, but it is not devoid of drawbacks as illustrated in Table 8.

A DDD scheme can be used not only to compare drug consumption in wards, hospitals, regions, or countries, but also across different time periods, as we did in the present study. A calculation of DDD/100 or 1000 patient-days enables the exclusion of differences stemming from the different ward bed count and occupancy. Thus, DDD-assessed antibiotic consumption is a reliable tool to compare drug consumption between similar wards (Sabaté et al. 2015; Muller et al. 2006). However, the value of DDD/100 hospitalizations or DDD/100 patient-days is influenced by many factors, e.g., the referral level of the hospital, the incidence of infectious diseases in the studied period, the epidemic foci in the hospital, the prescription habits in the ward, the promotion of certain drugs, and

Table 7 Drug use rates (DU100% and DU90%) before (2011) and after (2012–2016) the implementation of the hospital antibiotic policy

Indicator (per year)	2011	2012	2013	2014	2015	2016
DU100%	6: Ceftriaxone Ciprofloxacin Clindamycin Amoxicillin + clavulanate Imipenem Cefuroxime	7: Amoxicillin + clavulanate Cefuroxime Ceftriaxone Clindamycin Ciprofloxacin Gentamicin Meropenem	7: Amoxicillin + clavulanate Gentamicin Cefuroxime Ceftriaxone Ciprofloxacin Amoxicillin Amikacin Clindamycin	11: Ciprofloxacin Amoxicillin + clavulanate Cephazolin Ceftriaxone Cefuroxime Piperacillin + tazobactam Clindamycin Doxycycline Imipenem Vancomycin Amikacin	5: Ciprofloxacin Amoxicillin + clavulanate Amikacin Cephazolin Ceftriaxone	12: Ciprofloxacin Amoxicillin + clavulanate Cephazolin Ceftriaxone Cefuroxime Piperacillin + tazobactam Clindamycin Meropenem Vancomycin Amikacin Azythromycin
DU 90%	2: Amoxicillin + clavulanate Cefuroxime	3: Amoxicillin + clavulanate Cefuroxime Ceftriaxone	5: Amoxicillin + clavulanate Gentamicin Cefuroxime Ceftriaxone Ciprofloxacin	5: Ciprofloxacin Amoxicillin + clavulanate Cefuroxime Clindamycin Amikacin	3: Ciprofloxacin Amoxicillin + clavulanate Amikacin	3: Ciprofloxacin Amoxicillin + clavulanate Amikacin

Table 8 Strengths and limitations of defined daily doses (DDD) according to Muller et al. (2006)

Strengths of DDD	Limitations of DDD
Lack of dependence on drug cost and package size	Wards treating children cannot be compared to those treating adults
Consumption patterns of drugs can be compared throughout the country and internationally, at various time intervals in the hospital or ward; providing that hospital wards of similar profiles were compared with each other.	DDD may be different from the actual dose the patient receives, e.g., dose adjustment in patients with renal failure, dose calculated to body mass, or dosing depending on pharmacokinetics and patient's compliance
	DDD is not necessarily the dose recommended by the manufacturer, e.g., depending on the pathogen or severity of infection

even the season of year. Thus, assessment should be treated with caution, taking note of sudden and large increases in DDD/100 patients-days and of therapy costs of a single person-day, which we considered in the present study. Moreover, our assessment concerned the consumed DDD and not the hospital pharmacy data that do not take into account the number of expired and damaged drugs or the surplus of drugs in the ward. We conclude that there is an obvious need for the continuous evaluation and monitoring of the hospital antibiotic policy.

Conflicts of Interest The authors declare no conflict of interests in relation to this article.

References

Ayazi K, Khabaz A, Ayazi L, Ghorbani B, Eslami M, Ebrahimi M (2015) Changes in antibiotic use in a general surgery unit over a 5-year period. East Mediterr Health J 21:134–139

Bao L, Peng R, Wang Y, Ma R, Ren X, Meng W (2015) Significant reduction of antibiotic consumption and patients' costs after an action plan in China, 2010–2014. PLoS One 13(10):e0118868

Birger RB, Kouyos RD, Cohen T, Griffiths EC, Huijben S, Mina MJ, Volkova V, Grenfell B, Metcalf CJE (2015) The potential impact of coinfection on antimicrobial chemotherapy and drug resistance. Trends Microbiol 23(9):537–544

Davey P, Brown E, Fenelon L, Finch R, Gould I, Hartman G, Holmes A, Ramsay C, Taylor E, Wilcox M, Wiffen P (2005) Interventions to improve antibiotic prescribing practices for hospital inpatients. Cochrane Datebase Syst Rev 4:CD003543

Davies J, Davies D (2010) Origins and evaluation of antibiotic resistance. Microbiol Mol Biol Rev 74:417–433

Dik JW, Vemer P, Friedrich A, Hendrix R, Lo-Ten Foe JR, Sinha B, Postma MJ (2015) Financial evaluations of antibiotic stewardship programs – a systematic review. Front Microbiol 6:546–555

Dumarin C, L'Heriteau F, Pefau M (2010) Antibiotic use in 530 French hospitals: results from a surveillance network at hospital and ward levels in 2007. J Antimicrob Chemother 65:2028–2036

Hammuda A, Hayder S, Elazzazy S, Black E (2013) Point prevalence survey of antimicrobial utilization in oncology patients. J Infect Dev Ctries 7:990–993

Kim JY, Sohn JW, Park DW, Yoon YK, Kim YM, Kim MJ (2008) Control of extended-spectrum beta-lactamase-producing *Klebsiella pneumoniae* using a computer-assisted management program to restrict third-generation cephalosporin use. J Antimicrob Chemother 62:416–421

Mascarello M, Simonetti O, Knezevich A, Carniel LI, Monticelli J, Busetti M, Schincariol P, Torelli L, Luzzati R (2017) Correlation between antibiotic consumption and resistance of bloodstream bacteria in a University Hospital in North Eastern Italy, 2008–2014. Infection 45(4):459–467

Miliani K, L'Hériteau F, Alfandari S, Arnaud I, Costa Y, Delière E (2008) Antimicrobial surveillance network. Specific control measures for antibiotic prescription are related to lower consumption in hospitals: results from a French multicentre pilot study. J Antimicrob Chemother 62:823–829

Muller A, Monnet DL, Talon D, Hénon T, Bertrand X (2006) Discrepancies between prescribed daily doses and WHO defined daily doses of antibacterials at a university hospital. Br J Clin Pharmacol 61:585–591

Nitsch-Osuch A, Kurpas D, Kuchar E, Zycińska K, Zielonka T, Wardyn K (2015a) Antibiotic consumption pattern in the neonatal special care unit before and after implementation of the hospital's antibiotic policy. Adv Exp Med Biol 835:45–51

Nitsch-Osuch A, Kuchar E, Życińska K, Gyrczuk E, Miśkiewicz K, Korzeniewski K (2015b) Implementation of hospital's antibiotic policy decreases antimicrobial use in the general pediatric ward. Adv Exp Med Biol 857:67–74

Roca I, Akova M, Baquero F, Carlet J, Cavaleri M, Coenen S (2015) The global threat of antimicrobial resistance: science for intervention. New Microbes New Infect 6:22–29

Sabaté M, Ferrer P, Ballarín E, Rottenkolber M, Amelio J, Schmiedl S (2015) Drug utilization in Europe: nationwide data sources and a review of publications on a selected group of medicines (PROTECT project). Basic Clin Pharmacol Toxicol 116:201–211

Schelenz A, Nwaka D, Hunter P (2015) Longitudinal surveillance of bacteraemia in haematology and oncology patients at a UK cancer centre and the impact of ciprofloksacin use on antimicrobial resistance. J Antimicrob Chemother 12:1431–1438

Schwaber M, Carmelli Y (2008) Carbapenem-resistant Enterobacteriaceae: a potential threat. JAMA 300:2911–2913

Sullivan T (2014) Antibiotic overuse and *Clostridium difficile*: a teachable moment. JAMA Intern Med 174:1219–1220

Urbánek K, Kolár M, Lovecková Y, Strojil J, Santavá L (2007) Influence of third-generation cephalosporin utilization on the occurrence of ESBL-positive *Klebsiella pneumoniae* strains. J Clin Pharm Ther 32:403–408

Weichman KE, Levine SM, Wilson SC, Choi M, Karp NS (2013) Antibiotic selection for the treatment of infectious complications of implant-based breast reconstruction. Ann Plast Surg 71:140–143

Zilberberg MD, Nathanson BH, Sulham K, Fan W, Shorr AF (2017) Carbapenem resistance, inappropriate empiric treatment and outcomes among patients hospitalized with *Enterobacteriaceae* urinary tract infection, pneumonia and sepsis. BMC Infect Dis 17:279

Advs Exp. Medicine, Biology - Neuroscience and Respiration (2018) 38: 31–40
DOI 10.1007/5584_2017_121

Published online: 11 March 2018

Surgical Correction of Pectus Excavatum by the Nuss Procedure: A 15-Year Experience Study

Piotr Cierpikowski, Adam Rzechonek, Piotr Błasiak, Hanna Lisowska, Grzegorz Pniewski, and Patric Le Pivert

Abstract

Pectus excavatum is the most common congenital deformity of the chest. The Nuss procedure is minimally invasive surgical correction of this defect, using retrosternal metal bars. The purpose of the present study was to describe a 15-year experience with the Nuss surgery, and to evaluate the long-term clinical results of the procedure. We retrospectively evaluated 239 patients, aged 14–34, who underwent the Nuss surgery in the years 2002–2016. Postoperative complications were observed in 40/236 (16.9%) patients. The most common complication was pneumothorax in 14/239 patients. Less common were the following: wound infection in 4, pleural effusion in 3, allergy to nickel in 1, lung atelectasis in 1, and ventricular failure in 1 patient. Three patients were treated because of severe postoperative pain, and in one case the implant had to be removed. Postoperative complications associated with the number of bars inserted, but not with the patient age or gender. A satisfactory and long lasting corrective effect of surgery was observed in 231/239 (96.7%) of patients. There was no perioperative mortality. We conclude that the Nuss surgery is a safe surgery that demonstrates excellent and long-lasting esthetic results, with a low risk of severe complications.

Keywords

Chest wall · Congenital deformity · Nuss procedure · Pectus excavatum · Pneumothorax · Postoperative complications

1 Introduction

Pectus excavatum is the most common anterior chest wall deformity, which is characterized by conical depression of the sternum and costal cartilages. The incidence is estimated at 1–8:1000 births and it is 4–6 times more frequent in males (Cobben et al. 2014; Obermeyer and Goretsky 2012). Surgical treatment of pectus excavatum is due mainly to psychosocial issues stemming from the abnormal physical appearance. However, in some cases this defect may

P. Cierpikowski, A. Rzechonek (✉), and P. Błasiak
Department of Thoracic Surgery, Wrocław Medical University, Wroclaw, Poland

H. Lisowska and G. Pniewski
Department of Thoracic Surgery, Lower Silesian Center for Lung Diseases, Wroclaw, Poland

P. Le Pivert
Interventional Drug Delivery Systems and Strategies Group, Jupiter, FL, USA

result in decreased exercise tolerance, chest pain, or a severe cardiopulmonary dysfunction (Lawson et al. 2011; Malek et al. 2003).

The history of surgical correction of pectus excavatum started at the beginning of twentieth century, but the first techniques were unsatisfactory. In 1949, Mark Rawitch introduced a new open method that required a subperichondrial resection of deformed cartilages, osteotomy of the sternum, and suturing through the sternal bone to maintain a new shape (Rawitch 1949). That procedure quickly gained popularity and was used by physicians for several decades. However, the Rawitch method was characterized by serious invasiveness and a high risk of recurrence. A milestone in the development of surgical treatment of pectus excavatum was the introduction of a new minimally invasive procedure by Donald Nuss in 1998, which consisted of remodeling of the anterior chest wall by employing a retrosternal metal bars, with avoidance of cartilage resection (Nuss et al. 1998). The elasticity of the chest wall is a key factor in this procedure. The process of cartilage depletion ends completely by age 18–20 years, so that treatment after this age becomes more difficult than that in children and adolescents. Currently, the Nuss procedure has become the method of choice for patients with pectus excavatum (Zhang et al. 2015; Kelly et al. 2010).

The aim of the present study was to describe our 15-year experience with the Nuss procedure. We set out to evaluate complications and failures associated with this surgery, paying particular attention to the influence of patient age and the shape and degree of deformity. We further addressed the adverse effects having to do with the poor perception of treatment outcomes by the patient and attempted to demonstrate the ways to prevent and treat surgery failures.

2 Methods

2.1 Patients

This study was performed in accord with the Declaration of Helsinki for Human Research of the World Medical Association and it was approved by the Bioethics Committee of Wroclaw Medical University in Wroclaw, Poland. The study is retrospective in nature as it is based on the information collected from the patient medical files. It included 236 patients diagnosed with pectus excavatum, including eight patients with a recurrence of pectus excavatum deformity, who were treated surgically by the Nuss procedure in the Department of Thoracic Surgery of the Wrocław Medical University in Wroclaw, Poland in the years 2002–2016. There were 197 (83.5%) men and 39 (16.5%) women. The male-to-female ratio was 5 to 1. Patients' age ranged from 14 to 34 years (the mean of 17.4 ± 3.8 years). All patients were preoperatively examined. A routine examination consisted of basic laboratory tests, chest X-ray in posterior-anterior and lateral projections, spirometry, and an electrocardiogram. In selected cases, computed tomography was carried out. The majority of patients who were qualified for the Nuss procedure reported the cosmetic reasons as the main cause of a decision to undergo a surgical correction. The patients were subdivided into two groups: Group A – children and adolescents, and Group B adults. The cut-off age for this separation was 18 years. Groups A and B did not differ in terms of gender, quality of deformation, and the number of implants used. Details of patient characteristics are displayed in Table 1.

Chest deformities were distinguished by the presence of symmetry or lack thereof. The need to use one or two implants was taken as defining the extent of deformation. The majority of corrected deformations were symmetric (Fig. 1). However, chest asymmetry in the sagittal axis was also subject to correction in seven (3%) patients (Fig. 2). Serious distortions in the long sternal axis, with the tumefaction of the sternum manubrium, twisting the sternum over 30–35° or corners of the sterno-costal joint, were corrected by the Ravith procedure. The evaluation of the present study takes into consideration the reasons for early, occurring up to 30 days, postsurgery complications, and all subjective failures in the cosmetic aspects. We further

Table 1 Patient characteristics

	Group A (n = 149)	Group B (n = 87)	p-value	Total n = 236
Gender				
Female	25 (16.8%)	14 (16.1%)	ns	39 (16.5%)
Male	124 (83.2%)	73 (83.9%)	ns	197 (83.5%)
Age (years)				
Mean	15.7 ±1.7	22.3 ± 3.9	ns	–
Median	16.0	21	ns	–
Distortion type				
Symmetric	143	86	ns	229
Asymmetric	6	1	ns	7
Number of bars				
One bar	145 (97.3%)	80 (92.0%)	ns	225 (95.3%)
Two bars	4 (2.7%)	7 (8.0%)	0.059	11 (4.7%)

Differences between Group A and Group B
ns nonsignificant (Chi^2 test)

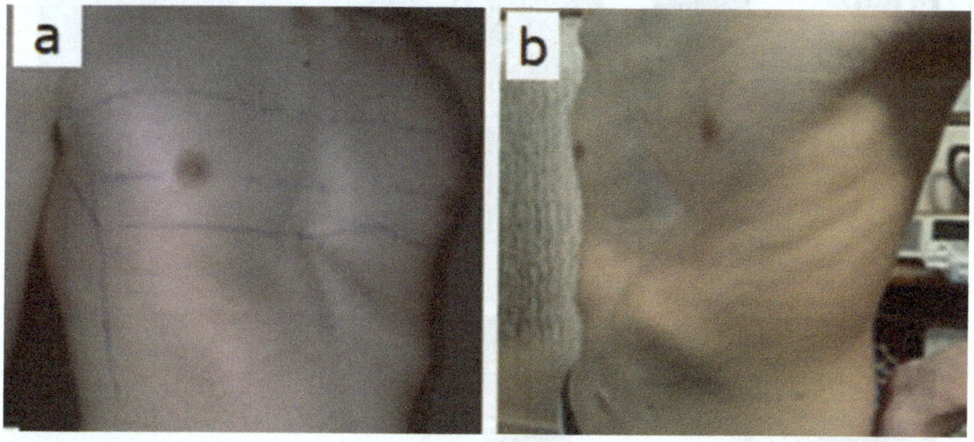

Fig. 1 Chest deformations: (**a**) symmetric, confined to the sternum body, (**b**) symmetric of the 'canyon' type, extending over the entire sternum

collected detailed medical history consisting of the type of surgical correction, the length of hospitalization, the number of bars used, the clinical course of treatment, and the rate of recurrence.

2.2 Surgical and Postsurgical Interventions

When a defect was visible in the long axis of the sternum, consisting of a depression in the sternum body, hilt of manubrium deformation, sternum twisting over 45°, presence of bumps in the sterno-costal joint, or a large deformation of xiphoid process, the patients were qualified for the Rawitch method. All surgeries were performed under general anesthesia, with tracheal intubation. Figure 3 demonstrates the insertion procedure. Two incisions were made bilaterally at the height of a lower part of the sternum and a guideway was created under the sternum. A measure of a preshaped metal bar was taken for insertion (Fig. 3a). The bar was inserted, semicircular bent directed downward, retrosternally in line with the deepest point of depression (Fig. 3b). Then, the bar was rapidly and forcefully turned over 180° to shore up the sternum depression point (Fig. 3c), and the chest wall layers were sutured, with a drainage tube

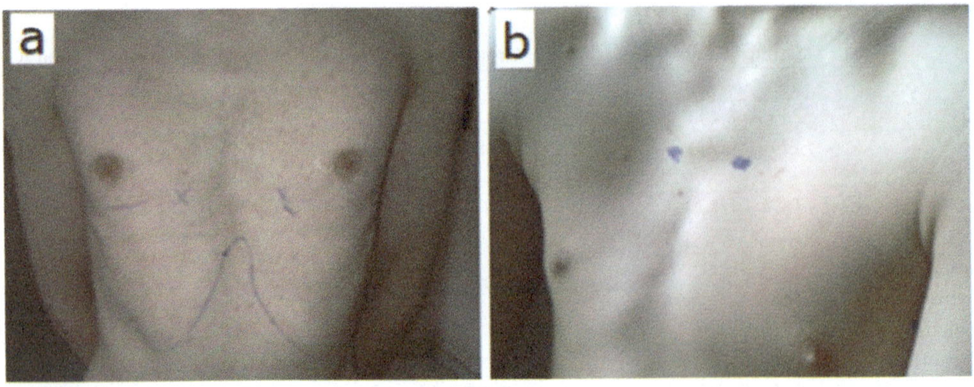

Fig. 2 Chest deformations: (**a**) asymmetric, confined to the sternum body, (**b**) asymmetric of the 'canyon' type

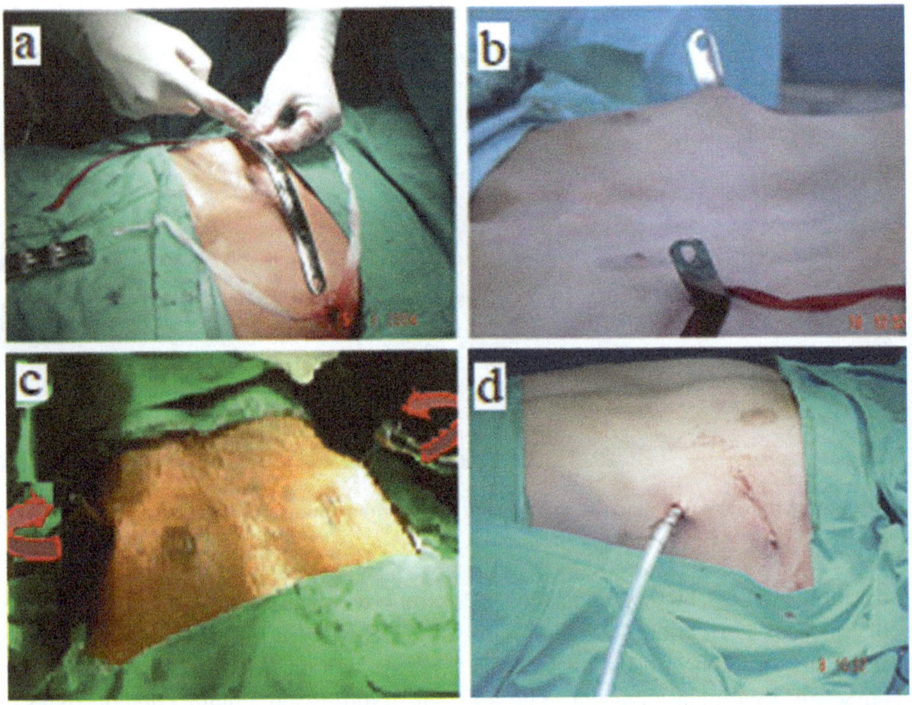

Fig. 3 The Nuss procedure for a metal bar implantation: (**a**) creation a channel guideway under the sternum and taking a measure of a preshaped metal bar for insertion, (**b**) insertion of the bar retrosternally, (**c**) forceful rotation of the bar (red arrows) into the correction position to shore up the deformation and (**d**) suturing and drainage of pleural cavity

attached (Fig. 3d). The procedures were carried out under a video-thoracoscopic guidance. In the majority of 236 patients (95.3%) we used one metal bar to obtain a satisfactory correction, but in a few severe and asymmetric cases of chest deformity two metal bars were required (Table 1).

Antibiotics were used preoperatively. In the postoperative period, paravertebral anesthesia with 0.25% bupivacaine was routinely used.

Most patients were discharged 5–7 days after surgery and were followed-up regularly. The patients were advised to avoid bending, sports, heavy physical activity, or carrying heavy objects to prevent the displacement of bar. Implants were maintained, on average, for 2 years. Then, the bars were removed under general anesthesia through incision along the previous scar.

2.3 Outcome Evaluation

The patients were followed up at 1, 3, 6, 12, and 24 months after surgery. Every complication during treatment or in a postoperative period was reported in the medical history. The results of surgical correction were summarized after removal of bars. A normal position of the sternum after removal of implants was considered a satisfactory result. There was then no need of any additional surgical correction and the patient expressed the acceptance of the final treatment result using a specially designed questionnaire. An unsatisfactory effect of treatment was considered when the deformation recurred after removal of bars, or when there was severe discomfort, pain, or the explicit patient's dissatisfaction.

Categorical clinical data between the two groups were compared with a Chi squared test. A p-value <0.05 was assumed to define statistically significant differences. Statistical evaluation was performed using Statistica PL v12.0 (StatSoft; Cracow, Poland).

3 Results

The final effect of the Nuss procedure was assessed as satisfactory in 227 (96.2%) of patients. In another nine patients, additional corrective procedures had to be performed to achieve a satisfactory result, due to deformity recurrence in 8 patients and a sternum fracture in one patient. The mean length of hospitalization was 8.3 days. There has been no perioperative mortality. The long-term complications were observed in 40 (16.9%) patients. Complications occurred significantly more often in the patients who required the insertion of two bars due to a severe chest deformity ($p < 0.001$). The frequency of complications was unrelated to age or gender. Postsurgical complications in patients operated on because of the presence of pectus excavatum are displayed in detail in Table 2.

The most common complication was pneumothorax that occurred in 14 patients (Table 3). These patients were qualified for pleural drainage or were conservatively treated. In three patients, there were seromatous pleural effusions in the postoperative wound. Hemothorax occurred in one patient, which necessitated a redrainage. Wound infection occurred in 4 (10%) cases of complications and responded to antibiotic treatment. Three patients suffered from severe postoperative chest pain that required additional surgical interventions. In two of those patients, intercostal nerve block sufficed to relieve pain, but another patient required surgical removal of a metal bar to relieve neuropathic pain. Bar displacement was noted in three patients. One of those patients required additional stabilizing attachment of the implant to the sternum, and the other two had to undergo a bar removal. We also observed a case allergy to the nickel component of a bar. The implant had to be replaced by metal bar without nickel. In one patient, there was lung atelectasis that was treated by bronchial aspiration and breathing exercise. We also had one case of severe right ventricular failure as a result of pericardial injury cause by the implant during the surgical procedure. A fracture of the sternum occurred in one patient (Fig. 4). The fracture was caused by the inadequate distribution of forces acting on the anterior chest wall, and it required reoperation. We did not observe any statistically significant differences in the frequency of postoperative complication between the adolescent and adult groups.

Overall, additional surgical interventions were performed in 9 (3.8%) patients. The interventions consisted of 13 reoperations and one intraoperative correction. The main cause for additional surgery was a recurrence of deformity. The most common intervention was the removal of a metal bar which

Table 2 Association between the presence of complications and perioperative parameters

Parameter	Postoperative period No complications; n (%)	With complications; n (%)	p-value
Age (years)			
<18	125 (83.9)	24 (16.1)	ns
≥18	72 (82.8)	15 (17.2)	
Gender			
Female	35 (89.7)	4 (10.3)	ns
Male	162 (82.2)	35 (17.8)	
Number of bars			
One bar	193 (85.8)	32 (14.2)	<0.001
Two bars	4 (36.4)	7 (63.6)	

Number of cases (%)
ns nonsignificant

Table 3 Character of postsurgical complication in patients with pectus excavates in adolescents (Group A) and adults (Group B)

Complication	Group A (n = 24) *n (%)*	Group B (n = 16) *n (%)*	p-value	Total (n = 40) *n (%)*
Pneumothorax – resorption	7 (29.2)	0	0.021	7 (17.0)
Pneumothorax – drainagae	4 (16.7)	3 (19.0)	ns	7 (17.0)
Pleural effusion	2 (8.3)	1 (6.0)	ns	3 (7.5)
Hemothorax	0	1 (6.0)	ns	1 (2.5)
Wound infection	3 (12.5)	1 (6.0)	ns	4 (10.0)
Intercostal neuralgia	1 (4.2)	2 (13.0)	ns	3 (7.5)
Bar displacement	1 (4.2)	2 (13.0)	ns	3 (7.5)
Nickel allergy	0	1 (6.0)	ns	1 (2.5)
Atelectasis	1 (4.2)	0	ns	1 (2.5)
Heart failure	1 (4.2)	0	ns	1 (2.5)
Sternum fracture	1 (4.2)	0	ns	1 (2.5)
Recurrence of deformity	3 (12.5)	5 (31.0)	ns	8 (20.0)

Number of cases (%)
ns nonsignificant

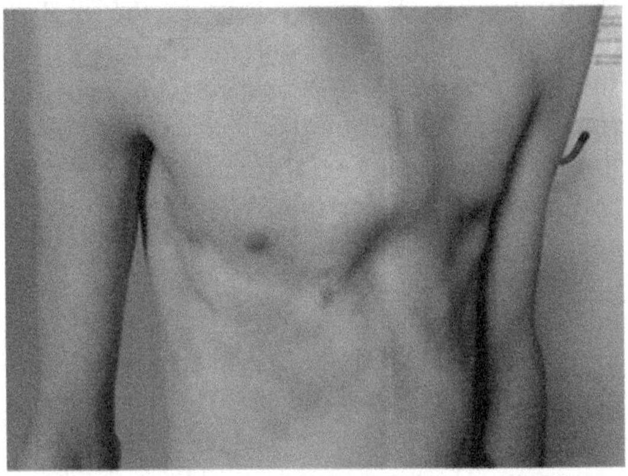

Fig. 4 Surgical complication after employing the Nuss procedure; a fracture of the sternum at the implant level

was done in five cases. In four patients, deformity was reoperated by the Rawitch method. A recurrence of deformation was not observed in patients below 16 years of age.

4 Discussion

Cosmetic effects after corrective thoracic cage surgery using the Nuss procedure are quite satisfactory. According to various authors, the surgery is rated as success in over 90% of cases (Park et al. 2004; Dzielicki et al. 2006). The complications that arise after surgery are transient, reversible, and non-life threatening, even though some of them may be cumbersome to the patient. The percantage of complications and failures noted in the present study was 16.3%, which is comparable to those listed in other studies performed in the groups of patients numbering more than 200 subjects of varied age, such as 16.1% (Park et al. 2004), 18.7% (Umuroglu et al. 2013), or 35.0% (Pawlak et al. 2016). The frequency of complications decreases with the acquisition of surgery experience.

Qualification of a given occurrence as a surgery complication or failure depends on the subjective evaluation of an author. That might be exemplified by a study of Pawlak et al. (2016) describing a large group of 680 regularly monitored patients, in which, apart from the 35% complication rate, carefully collected information on all cosmetic effects is listed. A comparable 37% complication and failure rate has been reported in an earlier study of Antonoff et al. (2009). Different qualification criteria could account for a relatively low incidence (6.4%) of reported complications in a multicenter study, covering a group of 777 patients in Saxena et al.'s (1999) metaanalysis.

The most frequently described and relatively easy to control early complications are the pleural disorders of pneumothorax and effusion. In the present study, pneumothorax accounted for 34% of postsurgery complications, and it required drainage only in one half of the instances. Small air sacks, located at the top of the pleura, underwent a spontaneous resolution. There is a wide spread of data in the literature on the occurrence of pneumothorax, ranging from 4.2% to 16.1% of cases (Umuroglu et al. 2013; Wu et al. 2016; Park et al. 2004; Jacobs et al. 2002). Sometimes, pneumothorax has not considered as a complication at all (Kelly et al. 2013). Pleural effusion has been reported in 2.5% to 10% of cases (Jo et al. 2003; Schaarschmidt et al. 2002). In the present study, small pleural effusions in the diaphragm angle appeared only in three cases, corresponding to 7.5% of all complications. Intrapleural bleeding or pleural hematoma requires surgical intervention. There was one such case that required pleural drainage. The literature describes single cases treated with thoraco-centesis (Park et al. 2004) or requiring thoracotomy (Bilgi et al. 2017). Pleural complications can be avoided, to an extent, by leaving a thin catheter at the top of the pleural cavity for 2–3 days after surgery.

The displacement of a correction plate can be a serious complication of the Nuss procedure, with the frequency ranging from 1% to 33% of cases as described in various studies (Wu et al. 2016; Kelly et al. 2013; Park et al. 2004; Fonkalsrud 2003; Hebra et al. 2001; Engum et al. 2000). To ensure a metal bar stability, recommend to avoid the use of absorbable stitches for the implant fixation. It also is recommended to place one suture to a rib in the parietal region. Schaarschmidt et al's (2002) approach with 14 absorbable figure of 8 sutures around the bar placed directly on the ribs in a submuscular position gives some protection against the bar displacement. In the present study, bar displacement took place in three cases, constituting 7.5% of complications. In one of those cases, it was caused by the erroneous repeat rotations of the bar during its fitting to the chest. Another serious complication of the surgery was a wider than necessary tunnel for two implants positioned too close to each other. Yet another case of implant displacement was a failure of the patient to follow the postoperative recommendations.

Although operative access in the Nuss procedure is achieved through small incisions, local complications occur in the surgical wound. These complications are often described as a serous effusion, which occurs in 1% to 6.5% of cases (Jo et al. 2003). In the present study, the effusion occurred in 3 cases, which constitutes 7.5% of all complications observed. All were treated with locally, without having to remove the implant. Some of the wound exudates could result from *a priori* unrecognized allergic reaction to nickel. In the present study, in most cases, no allergy tests on metals were performed. We observed postsurgery allergic reactions in one patients, i.e., 2.5% of all complications. The Nuss surgical procedure afflicts pain, reported in up to 5.0% of cases, which is usually greater than that after the Rawitch or Leonard procedure (Antonoff et al. 2009). It becomes a complication requiring intervention if it is not noticeably reduced after the perioperative period (Hebra et al. 2001; Engum et al. 2000). In the present study, pain persisted over 30 days after surgery in three patients; 7.5% of all complications. In two of those cases, pain was controlled by intercostal nerve block and analgesics. In one case, neuropathic pain required reoperation with removal of the implant. In addition to proper postoperative treatment, intraoperative local analgesia may be an important preventive measure against pain. Also, the use of a shorter implant, with slight retraction and bent ends, prevents a later compression of periosteal ribs, which may be a source of pain.

A failure to obtain the satisfactory cosmetic appeasement does not belong to complications, but it may determine the overall deterioration of surgery effects. The assessment of the chest appearance after correction is subjective, most often independent of the dimension of the changes made, and it is related to the preferences and psychic characteristics of patients. Therefore, in the present study, attention was paid only to the most disadvantageous cosmetic effects that prompted the patient and surgeon for reoperation or for other corrective measures. We qualified patients for the Nuss surgery mostly those who had symmetric deformities. The creator and promoter of the method has reported very good results in 71%, good in 21%, and a relapse in 7.8% of patients (Nuss et al. 2002). Fonkalsrud (2003), comparing the long-term results of the Nuss surgery with other methods, has reported that the need for reoperation occurred after minimally invasive surgery in six patients, and in none operated with the modified Rawitch surgery. Hosie et al. (2002) have reported that the early results are very good or good in 81.5% of cases, but distant results have not been provided. In several cases, we also qualified for surgery patients with asymmetry in the sagittal axis of the sternum. However, in these severely distorted cases, such as a fusion of the sternum cavity with the manubrium bulge, twisting of the sternum over 45°, or horns at the sterno-costal joint, we employed the Rawitch method of surgery in line with Cartoski et al.'s (2006) recommendation. That may have a bearing on a relatively low number of failures and recurrences we encountered. Park et al. (2004) have suggested the presence of a 'grand canyon' type of deformation, along with an inadequate initial experience of the surgical team in treating this kind of deformation, as a cause of the high 16% rate of complications. However, another reason for the complications reported in that study could be the high 43% rate of asymmetric distortions qualified for the Nuss surgery.

In the present study, the need for reoperation occurred in eight patients, accounting for 20% of all complications. We found no difference in the incidence of complications between adolescents and adults. Likewise, there were no age-related differences in the frequency of deformity recurrence. Such findings have also been reported, albeit in a lower age range, in other studies. Nakagawa et al. (2008) have set the age threshold for comparison at 10 years of age. The computed tomographic preoperative/postoperative index fell, on average, from 6.20 ± 3.58 to 2.93 ± 0.49 in the group below 10 years of age and from 5.50 ± 2.64 to 3.40 ± 0.79 in the group over 10 years of age. The difference was statistically insignificant between the two age-groups, therefore, the authors conclude that the Nuss procedure may be used independent of the patient's age. That is somehow at variance with the findings of Johnson et al. (2014) who have shown more complications in persons over 17 years of age and of Pawlak et al. (2016) who have shown a greater perioperative morbidity in young patients, but a greater relapse rate in adults.

In the present study, the rate of postoperative complications was 17.2%, in adults and 16.1% in the group of children and adolescents. Kelly et al. (2010) and Lawson et al. (2011) have reported that surgery with one implant has similar effects in adults and children. Two implants are required in 24% of adult patients (Zhang et al. 2015). However, the use of two implants is essentially dictated by anatomical conditions rather than age (Johnson et al. 2014). In the present study, two implants were required in 11 patients; in four up to 15 years of age, and in seven older patients. The patients of the older group suffered from significantly more overall complications (Table 2).

The limitation of the present study is that we focused only on serious complications of the Nuss surgery, requiring surgical interventions. Another imperfection is the lack of a comparative spirometry assessment before and after surgery. There was no computed tomography performed for the accurate assessment of a degree of deformity on the Haller scale in patients treated before 2008. An additional limitation is a narrowing of the Nuss treatment to patients mainly with symmetric chest deformities confined to the sternum body.

5 Conclusions

Although the Nuss surgery is more commonly performed in children and adolescents than in adults, anatomical conditions rather than age, are a limitation of the method. The operation of the funnel chest deformity using the Nuss procedure is feasible in adults and it is more likely to require two implants. At this point, there is a greater chance of complications and failures. A careful selection of patients, with exclusion of almost all extensive and asymmetric deformation, is important for achieving good results in older patients. The best clinical effect is achieved in patients with symmetric distortions. When there is asymmetry, cosmetic results are sometimes worse, although usually acceptable by patients. The presence of asymmetric deformations is an indication to consider another surgical approach, for instance, the Rawitch method.

Acknowledgments This study was supported by the statutory budget of Wroclaw Medical University in Poland

Conflict of Interest The authors declare no conflicts of interest in relation to this article.

References

Antonoff M, Alexandra E, Erickson A, Hess D, Robert D, Acton R (2009) When patients choose: comparison of Nuss, Rawitch, and Leonard procedures for primary repair of pectus excavatum. J Pediatr Surg 44:1113–1119

Bilgi Z, Ermerak NO, Çetinkaya Ç, Laçin T, Yüksel M (2017) Risk of serious perioperative complications with removal of double bars following the Nuss procedure. Interact Cardiovasc Thorac Surg 24 (2):257–259

Cartoski MJ, Nuss D, Goretsky MJ, Proud VK, Croitoru DP, Gustin T, Mitchell K, Vasser E, Kelly RE Jr (2006) Classification of the dysmorphology of pectus excavatum. J Pediatr Surg 41(9):1573–1581

Cobben JM, Oostra RJ, van Dijk FS (2014) Pectus excavatum and carinatum. Eur J Med Genet 57 (8):414–417

Dzielicki J, Korlacki W, Janicka I, Dzielicka E (2006) Difficulties and limitations in minimally invasive repair of pectus excavatum – 6 years experiences with Nuss technique. Eur J Cardiothorac Surg 30 (5):801–804

Engum S, Rescorla F, West K, Rouse T, Scherer LR, Grosfeld J (2000) Is the grass greener? Early results of the Nuss procedure. J Pediatr Surg 35(2):246–251

Fonkalsrud EW (2003) Current management of pectus excavatum. World J Surg 27(5):502–508

Hebra A, Gauderer MW, Tagge EP, Adamson WT, Othersen HB Jr (2001) A simple technique for preventing bar displacement with the Nuss repair of pectus excavatum. J Pediatr Surg 36(8):1266–1268

Hosie S, Sitkiewicz T, Petersen C, Gobel P (2002) Minimally invasive repair of pectus excavatum – the Nuss procedure. A European multicentre experience. Eur J Pediatr Surg 12(4):235–238

Jacobs JP, Quintessenza JA, Morell VO, Botero LM, van Gelder HM, Tchervenkov CI (2002) Minimally invasive endoscopic repair of pectus excavatum. Eur J Cardiothorac Surg 21(5):869–873

Jo WM, Choi YH, Sohn YS, Kim HJ, Hwang JJ, Cho SJ (2003) Surgical treatment for pectus excavatum. J Korean Med Sci 18(3):360–364

Johnson WR, Fedor D, Singhal S (2014) Systematic review of surgical treatment techniques for adult and pediatric patients with pectus excavatum. J Cardiothorac Surg 9:25

Kelly RE Jr, Mellins RB, Shamberger RC, Mitchell KK, Lawson ML, Oldham KT, Azizkhan RG, Hebra AV, Nuss D, Goretsky MJ, Sharp RJ, Holcomb GW, Shim WK, Megison SM, Moss RL, Fecteau AH, Colombani PM, Cooper D, Bagley T, Quinn A, Moskowitz AB, Paulson JF (2013) Multicenter study of pectus excavatum, final report: complications, static/exercise pulmonary function, and anatomic outcomes. J Am Coll Surg 217(6):1080–1089

Kelly RE, Goretsky MJ, Obermeyer R, Kuhn MA, Redlinger R, Haney TS, Moskowitz A, Nuss D (2010) Twenty-one years of experience with minimally invasive repair of pectus excavatum by the Nuss procedure in 1215 patients. Ann Surg 252 (6):1072–1081

Lawson ML, Mellins RB, Paulson JF, Shamberger RC, Oldham K, Azizkhan RG, Hebra AV, Nuss D, Goretsky MJ, Sharp RJ, Holcomb GW, Shim WK, Megison SM, Moss RL, Fecteau AH, Colombani PM, Moskowitz AB, Hill J, Kelly RE Jr (2011) Increasing severity of pectus excavatum is associated with reduced pulmonary function. J Pediatr 159 (2):256–261

Malek MH, Fonkalsrud EW, Cooper CB (2003) Ventilatory and cardiovascular responses to exercise in patients with pectus excavatum. Chest 124 (3):870–882

Nakagawa Y, Uemura S, Nakaoka T, Yano T, Tanaka N (2008) Evaluation of the Nuss procedure using pre- and postoperative computed tomographic index. J Pediatr Surg 43(3):518–521

Nuss D, Kelly RE Jr, Croitoru DP, Katz ME (1998) A 10-year review of minimally invasive technique for the correction of pectus excavatum. J Pediatr Surg 33 (4):545–552

Nuss D, Croitoru DP, Kelly RE, Goretsky MJ, Nuss KJ, Gustin TS (2002) Review and discussion of the complications of minimally invasive pectus excavatum repair. Eur J Pediatr Surg 12(4):230–234

Obermeyer RJ, Goretsky MJ (2012) Chest wall deformities in pediatric surgery. Surg Clin North Am 92(3):669–684

Park HJ, Lee SY, Lee CS (2004) Complications associated with the Nuss procedure: analysis of risk factors and suggested measures for prevention of complications. J Pediatr Surg 39(3):391–395

Pawlak K, Gąsiorowski Ł, Gabryel P, Gałęcki B, Zieliński P, Dyszkiewicz W (2016) Early and late results of the Nuss procedure in surgical treatment of pectus excavatum in different age groups. Ann Thorac Surg 102(5):1711–1716

Rawitch MM (1949) The operative treatment of pectus excavatum. Ann Surg 129(4):429–444

Saxena AK, Schaarschmidt K, Schleef J, Morcate JJ, Willital GH (1999) Surgical correction of pectus excavatum: the Münster experience. Langenbeck's Arch Surg 384(2):187–193

Schaarschmidt K, Kolberg-Schwerdt A, Dimitrov G, Straubeta J (2002) Submuscular bar, multiple pericostal bar fixation, bilateral thoracoscopy: a modified Nuss repair in adolescents. J Pediatr Surg 37(9):1276–1280

Umuroglu T, Bostancı K, Thomas DT, Yuksel M, Gogus FY (2013) Perioperative anesthetic and surgical complications of the Nuss procedure. J Cardiothorac Vasc Anesth 27(3):436–440

Wu PC, Knauer EM, McGowan GE, Hight DW (2016) Repair of pectus excavatum deformities in children: a new perspective of treatment using minimal access surgical technique. Arch Surg 136(4):419–424

Zhang DK, Tang JM, Ben XS, Xie L, Zhou HY, Ye X, Zhou ZH, Shi RQ, Xiao P, Chen G (2015) Surgical correction of 639 pectus excavatum cases via the Nuss procedure. J Thorac Dis 7(9):1595–1605

Advs Exp. Medicine, Biology - Neuroscience and Respiration (2018) 38: 41–52
DOI 10.1007/5584_2017_129

Published online: 28 November 2017

Influence of Inhaled Corticosteroids on Bronchial Inflammation and Pulmonary Function in Chronic Obstructive Pulmonary Disease with Moderate Obstruction

Iza Toczyska, Ewa Zwolińska, and Andrzej Chcialowski

Abstract

Inflammation in the bronchial respiratory tract and lung parenchyma underlies the pathogenesis of chronic obstructive pulmonary disease (COPD). It consists of effector cell infiltration, changes in reticular basement membrane (RBM) thickness, and the content of inflammatory mediators. The aim of this study was to assess the influence of inhaled corticosteroids (ICS) on the number of inflammatory cells and RBM thickness in bronchial biopsies, and pulmonary function in patients with moderate COPD. Twenty four patients with newly diagnosed COPD were included into the study. Pulmonary function tests and fiber optic bronchoscopy with bronchial biopsies were performed before and after 12-month treatment in two groups: ICS– group (LABA plus anticholinergics) and ICS + group (LABA plus anticholinergics plus ICS). We found that the addition of inhaled corticosteroids to the therapeutic regimen contributed to a reduction of RBM thickness, inflammation, and lung hyperinflation. The intensity of bronchial inflammatory infiltration had little effect on lung function. In conclusion, RBM thickness, an airway wall remodeling element, does not significantly affect the degree of airflow limitation.

Keywords

Bronchial inflammation · Chronic obstructive pulmonary disease · Fiber optic bronchoscopy · Inhaled corticosteroids · Pulmonary function · Reticular basement membrane

1 Introduction

Chronic obstructive pulmonary disease (COPD) is one of the most important causes of morbidity and mortality in the world. According to WHO, it will be the fifth cause of disability and the third cause of mortality by 2020 all over the world

I. Toczyska
Department of Internal Diseases, Pneumology, Allergology and Clinical Immunology, Military Institute of Medicine, Warsaw, Poland

E. Zwolińska
Department of Gynecology, Holy Family Maternity Hospital, Warsaw, Poland

A. Chcialowski (✉)
Department of Infectious Diseases and Allergology, Military Institute of Medicine, Warsaw, Poland
e-mail: achcialowski@wim.mil.pl

(GOLD 2014; Lopez et al. 2006; Mathers and Loncor 2006). COPD is a slowly progressing chronic disease characterized by airflow limitation, which is in most cases irreversible, a decline in lung function over time, and is accompanied by respiratory symptoms, such as dyspnea, cough, and sputum production. Cigarette smoking is the main etiologic factor leading to bronchial obstruction. Other less common factors include environmental and working pollution, and alpha 1-antitrypsin deficiency (Turato et al. 1995; Celli et al. 2005). This gradual and progressive loss of lung function is a result of pulmonary emphysema due to destruction of lung parenchyma and inflammation-induced narrowing of small airways. Macrophages and T lymphocytes, especially the cytotoxic CD8 phenotype occurring in the epithelial and subepithelial layers, are the main cells found in a stable period of the disease (Petrescu et al. 2013; O'Donnell et al. 2006; Hogg 2004; Lams et al. 1998, 2000). During exacerbations, neutrophils, eosinophils, and mast cells are predominant and show a high functional stimulation to secrete many of inflammatory substances, including prostaglandins, thromboxanes, toxic oxygen radicals, multifunctional cytokines, and chemotactic factors such as growth-related oncogene-α (GRO-α), interleukin-6 (IL-6), IL-8, monocyte chemotactic protein-1 (MCP-1), macrophage inflammatory protein 1α (MIP-1α). These bioactive substances contribute to further inflow of cells and bronchial epithelial damage. The epithelium also actively contributes to the maintenance of inflammation, by synthesis and secretion of numerous substances, including some of the mentioned above and others such as granulocyte colony stimulating factor (G-CSF), tumor necrosis factor-α (TNF-α), and transforming growth factor-β (TGF-β) (GOLD 2014; Hurst et al. 2010).

Inflammation in COPD is characterized by the predominance of CD8+ lymphocytes in bronchiolar walls (Saetta et al. 1999; O'Shaughnessy et al. 1997). Studies of bronchial biopsies have shown hyperplasia and hypertrophy of mucous glands, atrophy of bronchial epithelium, or its metaplasia into stratified squamous epithelium. A loss of elastic elements of bronchial walls and lung parenchyma is characteristic, especially when emphysema develops. There is also an increase in the number of macrophages and eosinophils in bronchial subepithelium, and B cells and eosinophils in small airways. Neutrophils are observed in a higher percentage in induced sputum and bronchoalveolar lavage fluid (BALF), especially in patients with accelerated decline in lung function (Petrescu et al. 2013; Barnes 2004).

The functional symptoms in COPD are accompanied by morphological changes in the bronchial wall, which is known as remodeling, which consists of a complex of structural changes including fibrosis and bronchial smooth muscle hypertrophy, and secondary changes in bronchial and pulmonary arteries. Another anatomical structure is involved is the reticular basement membrane (RBM). RBM thickening and subepithelial fibrosis are characteristic for asthma and are associated with the deposition of laminin, fibronectin, proteoglycans, and type I, III, and V collagen (Sohal et al. 2010, 2011; Jeffery 2001). The reports on RBM thickening in COPD patients are contradictory. According to one study, the RBM is not thickened, while others have shown its thickening, fragmentation, and hyper-vascularity to variable extent (Soltani et al. 2010, 2012).

In the light of the current knowledge, participation of inflammatory cells and their mediators in the inflammatory process in COPD (6–10) is undeniable. On the other hand, there are only a few well documented studies on the effects of therapy, including inhaled corticosteroids (ICS), on the local and general inflammatory process, and on the breathing function, in COPD patients with moderate obstructions. Clinical effects of inhaled corticosteroids in COPD are less clear in comparison with asthma. Burge et al. (2000) have shown that the treatment with fluticasone propionate in moderate and severe COPD improves the forced expiratory volume in 1 s (FEV1) only during the first 3 month of treatment. It is believed that a more precise understanding of the local inflammatory mechanisms would lead to better preventive and therapeutic measures, aimed at counteracting the consequences of inflammation. Therefore, in the present study we attempted to assess the clinical relevance of local and systemic bronchial inflammation in newly diagnosed COPD patients with moderate obstruction. The study was performed in patients treated with long-acting β2 agonists (LABA) and cholinolytic (CH) drugs, who were followed by a period of 1 year. The patients were stratified into two subgroups distinguished by the presence or absence of additional treatment with corticosteroids: ICS– and ICS+ group, respectively. We assessed the degree of airflow obstruction and the intensity of local bronchial inflammation expressed by the cellular infiltrates in bronchial mucosa and the thickness of RBM.

2 Methods

2.1 Patients, Testing, and Instrumentation

The study was approved by the Ethics Committee of the Military Institute of Medicine in Warsaw, Poland. It was an open-label non-placebo-controlled trail that was primarily focused on the evaluation of morphological and spirometry changes. The presumption was that the results would only have to do with the action of drugs used. Twenty four patients, 19 men and 5 women aged 54–75 years, with newly diagnosed and untreated COPD with moderately obstruction, were enrolled into the study. The patients gave informed consent for the study procedures. In addition, each patient signed informed consent concerning the conduction of fiber optic bronchoscopy with bronchial wall biopsy. Patients' characteristic are displayed in Table 1.

COPD diagnosis was based on the following:

- chronic cough, expectorations, dyspnea, exposure to cigarette smoking;
- spirometry results:
- forced vital capacity (FVC) – % of predicted value;
- forced expiratory value in 1 s (FEV1) – % of predicted value;
- ratio FEV1/FVC persistently lower than the 5th percentile after a bronchodilator test; i.e., 20 min inhalation of the short acting β2 mimetic salbutamol (400 μg)
- lack of atopy: normal IgE serum level and a negative skin prick test for most aeroallergens such as dust mite, tree and grass pollen, cat and dogs epithelium.

Exclusion criteria were the following:

- active smoking
- cancer
- unstable coronary disease
- uncontrolled hypertension
- severe heart rhythm disturbances
- heart insufficiency

All patients underwent a functional breathing test, which included spirometry and body plethysmography using a Jaeger setup (Würzburg, Germany). The measurement of volumetric and flow volumes was performed during a forced exhalation maneuver recorded as flow-volume curve.

Plethysmography was carried out in a sealed cabin after the initial automatic calibration determining the relationship between volume and pressure change in the body cabin. Increasing the chest volume during inhalation, increased the cabin pressure. Vital capacity (VC) maneuver ended the measurement. Based on Boyle-Mariotte's law, which states that the product of pressure and volume of gas is constant, and Pascal's law of equal pressure at each point of the closed system, the pressure in the cabin and airways at the known volume of the cabin was calculated from the volume of gas. Other indicators included thoracic gas volume (TGV) at the end of exhalation, residual volume (RV), total lung capacity (TLC), RV%TLC, and airway resistance (Raw).

Bronchoscopy was performed using video bronchoscopy (BF 1 T160; Olympus Optical; Tokyo, Japan) of the outer diameter of 3.2 mm.

Table 1 Baseline characteristics of patients with chronic obstructive pulmonary disease (COPD) treated with the addition of inhaled corticosteroids (ICS+) and without corticosteroids (ICS–)

	Whole group	ICS+	ICS–
Sex M/F (n)	19/5	10/2	9/3
Age (years)	64.5 ± 10.5	66.3 ± 12.3	62.7 ± 8.7
FEV1% pred	61.6 ± 14.2	57.5 ± 16.7	64.8 ± 11.0
RV (ml)	167.8 ± 62.2	167.4 ± 37.4	168.1 ± 79.1
Raw (kPa/l/s)	120.2 ± 76.9	138.9 ± 91.1	105.1 ± 62.2
RV%TLC	53.5 ± 11.25	52.8 ± 11.5	54.0 ± 11.5
FVC% pred	105.9 ± 14.5	104.8 ± 17.2	107.2 ± 11.5

Local anesthesia was used with 2% xylocaine, followed by a short sedation with 1.0–2.5 mg dormicum (midazolam) i.v. (F. Hoffmann-La Roche; Basel, Switzerland). During macroscopic examination of the bronchial tree, the secretion was aspirated at a pressure of −0.02 MPa and sent for a microbiological evaluation. The fiberscope was then wedged in the middle lobe bronchus. Bronchoalveolar lavage was performed using three portions of 50 ml of isotonic NaCl solution, the volume of recovered fluid was about 90–100 ml, which accounted for 60–70% of the primary volume. Three to four endobronchial biopsies were taken from the segmental carinas of the right lower lobe bronchus. Specimens were subjected to morphological and histopathological examinations.

Spirometry, plethysmography, and bronchoscopy were repeated in like manner after 1 year of treatment.

2.2 Biopsy Specimen Processing and Cell Quantification

Biopsy specimens were embedded and snap-frozen −80% in liquid isopentane. They were then cut serially into series of 4-μm thick sections and were stored at −20 °C until use. Morphologically optimal tissue was selected from a preview of hematoxylin-eosin-stained slides. Sections were air-dried for 20 min and fixed in acetone for 10 min. Immunohistochemical staining was performed using monoclonal antibodies against the secreted form of eosinophil cationic protein (ECP) (EG2, 1:50; Sanbio; Uden, The Netherlands), neutrophil elastase (NP57, 1:40; Dako; Copenhagen, Denmark), CD3 (1:50; Becton-Dickinson; San Jose, CA), and CD68 (1:500; Dako) for the detection of eosinophils, neutrophils, lymphocytes, and macrophages, respectively. For each antibody, a total number of positively stained cells was counted to the depth of 100 μm below a randomly selected 1 mm^2 area of the epithelial basement membrane. The mean value was taken for the evaluation. The magnification of light microscopy was 400x to detect neutrophils, eosinophils, CD3, and CD68.

2.3 Reticular Basement Membrane (RBM)

RBM thickness was assessed on sections using light microscopy (Olympus; Tokyo, Japan) and computer aided image analysis. It was assessed five times in different areas of each section, with a minimum distance of 20 μm from each other. The mean thickness was taken into account.

2.4 Statistical Elaboration

Data were presented as means ±SD. Distribution of variables, checked with the Kolmogorov-Smirnov test, was skewed; therefore, the intergroup differences were compared with the Mann-Whitney U test. Differences between the corresponding variables after 12 months of treatment in the two groups of COPD patients were compared with the Wilcoxon rank-sum test. The relationship between variables was determined with Spearman's correlation coefficients. A p-value <0.05 defined statistically significant differences. The evaluation was performed with a commercial Statistica for Windows package (StatsSoft; Tulsa, OK).

3 Results

3.1 Lung Function Tests

Results of lung function tests are detailed in Table 2. The lowest FEV1 values of 57.8 ± 6.7%predicted were in the ICS + group, while they amounted to 64.7 ± 11.0%predicted in the ICS– group. After 1 year of treatment, FEV1 increased in both groups and was 60.0 ± 14.9%predicted in ICS+ and 71.1 ± 12.3%predicted in ICS–; the increase was significant in the latter group ($p < 0.05$). Baseline FVC was normal in both groups of patients and it remained so after 1 year of treatment. Likewise, RV was similar in both groups, with insignificant changes after treatment. Concerning RV%/TLC, higher values were observed in the ICS+ than ICS– group of patients both before and after treatment; the ratio decreased significantly in both groups after treatment. Raw was greater in both ICS groups before treatment, amounting to 139.0 kPa/l/s in the ICS+ group and 105.1 kPa/l/s in the ICS– group of patients. After 1 year of treatment, Raw significantly decreased by about 23% in the former group, only with a trend for reduction in the latter group. There were no significant differences in any of the variables recorded between the ICS– and ICS+ groups.

3.2 Histological Evaluation of Bronchial Mucosa

In the ICS+ group, the mean baseline RBM thickness was 7.5 ± 2.8 μm and it decreased to 5.6 ± 1.6 μm after 1 year of treatment. In ICS– group, The RBM also decreased significantly, albeit the decrease was somehow less. Nonetheless, there were no significant differences in the RBM thickness between both groups either before or after treatment. Nor was there any association between the membrane thickness and pulmonary function variables investigated in either group (Table 3).

3.3 Histological Evaluation of Bronchial Wall Biopsy Specimens

Histopathological examination of bronchial wall biopsy specimens at baseline and after 1 year of treatment showed a significant decrease in the total cell number, CD8+ T cells, and macrophages in both groups. The number of neutrophils significantly increased in the ICS– but not ICS+ group. The results are shown in Table 4.

3.4 Associations Between Cells Infiltrating Bronchial Wall and Pulmonary Function in COPD Patients

The most meaningful associations between the cells infiltrating the bronchial wall and the variables of pulmonary function tests in the ICS– group of COPD patients were as follows: the total number of cells associated with Raw both before ($r = 0.64$) and after treatment ($r = 0.60$), the number of macrophages before treatment inversely associated with RV both before ($r = -0.55$) and after treatment ($r = -0.63$), macrophages after treatment also inversely associated with RV after treatment ($r = -0.48$), neutrophils after treatment associated with RV after treatment ($r = 0.75$), and eosinophils after treatment inversely associated with FEV1 ($r = -0.72$) and TV%/TLC ($r = 0.56$) after treatment.

In the ICS+ group of COPD patients, the total number of cells associated with RV both before ($r = 0.71$) and after treatment ($r = 0.62$), and CD8+ lymphocytes associated with FEV1 before ($r = 0.63$) and with RV after treatment ($r = 0.76$). All the above outlined associations were significant ($p < 0.05$).

Table 2 Lung function tests in patients with chronic obstructive pulmonary disease (COPD) (n = 12 for both groups) before and after 12 months' treatment with the addition of inhaled corticosteroids (ICS+) and without corticosteroids (ICS–)

	FEV1		FVC		RV		Raw		RV%TLC	
Group	Before	After	Before	After	Before	After	Before	After	Before	After
ICS+ (***n* = *12***)	57.8 ± 6.7	60.0 ± 14.9	104.8 ± 17.3	109.7 ± 16.9	167.4 ± 37.4	168.3 ± 80.7	139.0 ± 91.1	109.2 ± 64.1	52.8 ± 11.5	48.0 ± 10.2*
ICS– (n = 12)	64.7 ± 11.0	71.1 ± 12.3*	107.2 ± 11.9	114.1 ± 14.9	168.1 ± 79.1	159.2 ± 7.8	105.1 ± 61.2	98.2 ± 59.8	54.0 ± 11.5	50.5 ± 12.8*

Values are mean %predicted ±SD unless otherwise indicated; *$p < 0.05$

Table 3 Baseline basement membrane thickness in patients with chronic obstructive pulmonary disease (COPD) ($n = 12$ for both groups) before and after treatment with the addition of inhaled corticosteroids (ICS+) and without corticosteroids (ICS–)

Group	Before treatment (μm)	After treatment (μm)
ICS+	7.5 ± 2.8	5.6 ± 1.6*
ICS–	6.6 ± 1.7	6.0 ± 1.2*

*$p < 0.05$

4 Discussion

COPD is the most common chronic respiratory disease, posing a significant health and epidemiological burden (GOLD 2014; Lopez et al. 2006; Mathers and Loncor 2006). A gradual loss of ventilatory reserve, assessed from the annual decrease in FEV1 is a characteristic feature of COPD. The pathogenic basis of this phenomenon lies in the inflammatory response to damaging factors, especially tobacco smoke. In this context, it is warranted to assess the possible pathological alterations that can occur under the course of treatment. The present study investigated the effect of add-on inhaled corticosteroid treatment of one-year-long duration on the cell infiltration of bronchial wall, the thickness of bronchial reticular basement membrane, and the lung function. Changes in the bronchial wall structure are an essential component of the lower respiratory tract remodeling, which is characteristic of bronchial asthma and also is the main factor responsible for limiting the reversibility of obstruction in COPD. Remodeling has likely to do with an abnormal bronchial epithelial repair process, with a significant interaction between epithelial and mesenchymal cells. As a consequence, there are increases in thickness of the bronchial basement membrane, subepithelial fibrosis with increased myofibroblast proliferation, and in deposition of collagen type I and III, fibronectin, and tenascin (Sohal et al. 2011; Amin et al. 2003; Jeffery 2001). The amelioration of the thickened bronchial basement membrane is a well-recognized effect of inhaled corticosteroid treatment in asthma patients. However, there is no consensus on the issue in COPD where the results are divergent, despite the remodeling being well-documented (Fabbri et al. 2003). Some studies fail to report the basement membrane thickening at all, others demonstrate a thickening that is less expressed than that in asthma, while yet others demonstrate a comparable degree of thickening in both diseases, but with differences in histological structure (Lieskier et al. 2009). Bourdin et al. (2007) have failed to substantiate any difference in the thickness and structure of the reticular layer of the basement membrane between the patients with mild asthma and COPD, but the basement membrane was significantly thicker in severe asthma patients. Those authors failed to find a relationship between the basement membrane thickness and FEV1 in asthma, regardless of disease severity, or in COPD. Such correlation appeared only when mild and severe asthma patients were combined for the analysis. In the present study, the basement membrane thickness before treatment was 7.5 ± 2.8 μm in the ICS+ group and 6.8 ± 1.7 μm in the ICS– group. After a year's treatment, a significant decrease in the basement membrane thickness appeared in the former compared with the latter group. We did not find any association of the basement membrane thickness with the variables of pulmonary function investigated. It appears, therefore, that airflow limitation and obstruction irreversibility may yet depend on other than basement membrane thickness mechanisms.

Apart from remodeling, airway obstruction is a consequence of local inflammation in different lung structures. This notion is supported by the present finding of an association between a total number of cell infiltrating the bronchial wall and Raw both before and after treatment in the ICS– group. The total number of cell infiltrating cells and Raw in this group of patients were lower than those in the ICS+ group both before and after treatment. Treatment significantly decreased the total number of cell infiltrating cells in both groups, along with a reduction in Raw, which however was significant only in the ICS+ group. We also found that total number of infiltrating cells was associated with RV both before and after treatment in the ICS+ group of patients, which puts into question the role of

Table 4 Inflammatory cells in bronchial biopsy before and after treatment in patients with chronic obstructive disease (n = 12 for both groups) with the addition of inhaled corticosteroids (ICS+) and without corticosteroids (ICS–)

	TCN		T lymophcytes		Macrophages		Neutrophils		Eosinophils	
Group	Before	After	Before	After	Before	After	Before	After	Before	After
ICS+ (n = 12)	48.3 ± 25.1	26.3 ± 18.4*	16.2 ± 12.1	5.2 ± 5.0*	11.5 ± 29.7	1.4 ± 1.8	1.0 ± 1.1	1.0 ± 1.1	2.3 ± 3.1	7.1 ± 19.6
ICS– (n = 12)	37.8 ± 27.9	22.2 ± 16.8**	16.7 ± 11.1	8.2 ± 6.2*	2.2 ± 1.3	1.8 ± 1.2	0.4 ± 0.7*	1.6 ± 3.0*	0.2 ± 0.4	0.4 ± 1.1

Data are mean ± SD of cells; * < 0.05, **p < 0.005

inhaled corticosteroid therapy in decreasing air trapping due to airflow limitation throughout the bronchial tree. We further found that a degree of airflow obstruction expressed by FEV1 inversely associated with the number of CD8+ T cells in both ICS– and ISC+ groups, which suggests that the intensity of respiratory tract inflammation due to accumulation of CD8+ lymphocyte, the predominant subpopulation of T cells, has a negative influence on FEV1. This association disappeared after treatment. Although CD8+ T cell infiltrates were significantly reduced in both ICS– and ISC+ groups after treatment, FEV1 improved significantly only in the ICS– group. Although a reduction of CD8+ T cells, and thus a local anti-inflammatory effect, tended to be more pronounced in the ICS+ group, the results reaffirm the suggestion that mechanisms other than just bronchial inflammation may be at play in the control of airway obstruction or air trapping in COPD. The presence of such mechanisms would extend the hitherto knowledge on the CD8+ lymphocyte-mediated bronchial inflammation as a predominant contributor to COPD progression. Olloquequi et al. (2010) have reported the increases in T and B lymphocytes in small bronchi and lung parenchyma in patients with moderate obstruction. Di Stefano et al. (2001) have reported that the density of CD8+ T cell infiltrates was lower in the small bronchioles of patients with severe obstruction compared with moderate forms of obstruction, which is explicable by smoking cessation in such patients and by treatment with inhaled corticosteroids. Nonetheless, CD8+ T cell distribution in lung parenchyma is fairly homogeneous and thus independent of the degree of anatomical lung injury. Those authors have found no correlation between CD8+ T cell counts in lung parenchyma and FEV1. A greater number of CD8+ T cells in lung parenchyma is a likely consequence of respiratory infections, commoner in patients with more advanced disease, which is conducive to CD8+ T cell recruitment.

In the present study, the number of macrophages, yet another component of inflammatory infiltration, failed to associate with the degree of obstruction. That is somehow at variance with other studies dealing with COPD pathogenesis, which have demonstrated such a relationship (O'Shaughnessy et al. 1997; Stanescu et al. 1996). Yet we found that macrophage count was greater in the ICS+ group, where the FEV1 was lower. After a year's treatment, the number of macrophages decreased in both groups, significantly more so in the ICS+ group, which was accompanied by an increase in FEV1 in both ICS+ and ICS– groups, albeit only in the latter group the FEV1 increase achieved statistical significance. These results, as was the case with CD8+ lymphocytes, also fail to confirm a relationship between the degree of obstruction and the severity of macrophage infiltration. Moreover, we found that the number of macrophages in the bronchial wall biopsy inversely associated with RV before and after the treatment, which however was the case only in ICS– group.

This association could be speculatively explained by the fact that peripheral airspace enhancement decreases the distribution density of cells. The degree of pulmonary hyperinflation would then be not a consequence of bronchial inflammation, but would only factor in the distribution of inflammatory cells. We found that in the ICS+ group at baseline RV tended to be lower, and with a significantly greater number of macrophages, compared with the ICS– group. After treatment, the RV value slightly increased, which was accompanied by a reduction in the number of macrophages, which seems in line with the presumed decrease in cell density due to the sheer airspace enhancement. Overall, these results suggest that treatment-dependent changes in the inflammatory cell count and RV value are more self-related than dependent on the presence or not of the add-on inhaled corticosteroid treatment.

Other inflammatory cells examined in the bronchial wall did not show any association with functional variables before treatment in either group. After treatment, the number of neutrophils and eosinophils in the bronchial biopsy specimens increased in both groups. There was a significant association of neutrophils with RV. Eosinophils associated with positively associated with RV%/TLC and negatively with

FEV1. The associations of neutrophils and eosinophils were present only in the ICS– group. Thus, it appears that steroid therapy failed to affect a desired reduction in both cell types, reportedly having to do with the severity of COPD. That could explain the unsatisfactory effects of inhaled corticosteroid therapy in COPD reported in many a study (Sohal et al. 2014; Calverley et al. 2007; Chanez et al. 2004; Burge et al. 2000; Hattotuwa et al. 2002; Culpitt et al. 1999; Cosio et al. 1998). In this context, it is worth mentioning a study of Górska et al. (2008) on the eosinophilic bronchitis in asthma and COPD, which demonstrates a high rate of eosinophils in a stable form of both diseases, and not necessarily in exacerbations as has been hitherto believed. In the present study we also noted an association between the membrane thickness and eosinophils accumulation in the baseline condition before treatment in the ICS+ group. After a year's therapy, a decrease, on average by 1.9 μm ($p = 0.02$), in the basemend membrane thickness was observed in this group, which was significantly greater than the corresponding 0.78 μm decrease in the ICS– group. These results indicate the basement membrane thickness was strongly associated with the intensity of eosinophilic inflammation in the bronchial mucosa before the commencement of inhaled corticosteroid therapy. The therapy, despite not affecting the level of local eosinophilia, could have a beneficial effect on the disease course by effectively decreasing the basement membrane thickness. Thus, targeting eosinophilic inflammation may be a new therapeutic goal in COPD.

To sum up, the present study attempted to assess the role of a long-term inhaled corticosteroid treatment on the course of COPD with moderate airway obstruction, considering the relationship between bronchial wall inflammation and lung function results. The major findings were as follows:

- reticular basement membrane thickness and inflammatory cells infiltrating of bronchial mucosa are the key elements in the assessment of inflammation;
- CD8+ lymphocyte infiltration of bronchial wall plays an essential role in airway obstruction determined by the level of FEV1;
- extent of bronchial cell infiltration has a little effect on lung function;
- it appeares that hyperinflation, rather than the intensification of bronchial inflammation, influences the distribution of inflammatory cells in COPD;
- inhaled corticosteroids treatment reduces reticular basement membrane thickness, local inflammation, and lung hyperinflation;
- reticular basement membrane thickness, as one of the remodeling elements, does not significantly affect the degree of airflow limitation.

Conflicts of Interest The authors declare no conflicts of interest in relation to this article.

References

Amin K, Ekberg-Jansson A, Lofdahl CG, Venge P (2003) Relationship between inflammatory cells and structural changes in the lungs of asymptomatic and never smokers: a biopsy study. Thorax 58(2):135–142

Barnes P (2004) Mediators of chronic obstructive pulmonary disease. Pharmacol Rev 56:515–548

Bourdin A, Neveu D, Vachier I, Paganin F, Godard P, Chanez P (2007) Specificity of basement membrane thickening severe asthma. J Allergy Clin Immunol 119 (6):1367–1374

Burge PS, Calverley PM, Jones PW, Spencer S, Anderson JA, Maslen TK (2000) Randomised, double blind, placebo controlled study of fluticasone propionate inpatients with moderate to severe chronic obstructive pulmonary disease: the ISOLDE trial. BMJ 320 (7245):1297–1303

Calverley PM, Anderson JA, Celli B, Ferguson GT, Jenkins CH, Paul W, Jones PW, Yates JC, Veslbo J (2007) Salmeterol and fluticasone propionate and survival in chronic obstructive pulmonary disease. N Engl J Med 356(5):775–789

Celli BR, Helbert RJ, Nordyke RJ, Schau B (2005) Airway obstruction in never smokers: results from the third National Health and nutrition examination survey. Am J Med 118(12):1364–1372

Chanez P, Bourdin A, Vachier I (2004) Effects of inhaled corticosteroids on pathology in asthma and chronic obstructive pulmonary disease. Proc Am Thorac Soc 1(3):184–190

Cosio MG, Ghezzo H, Hogg JC, Corbin R, Loveland M, Dosman J, Macklem PT (1998) The relations between structural changes in small airways and pulmonary-function tests. N Engl J Med 298(23):1277–1281

Culpitt SV, Maziak W, Loukidis S, Nightingale JA, Matthesws JL, Barnes PJ (1999) Effect of high dose inhaled steroid on cells, cytokines, and proteases in induced sputum in chronic obstructive pulmonary disease. Am J Respir Crit Care Med 160(5):1635–1639

Di Stefano A, Capelli A, Lusuardi M, Caramori G, Balbo P, Ioli F, Sacco S, Gnemmi I, Brun P, Adcock IM, Balbi B, Barnes PJ, Chung KF, Donner CF (2001) Decreased T lymphocyte infiltration in bronchial biopsies of subjects with severe chronic obstructive pulmonary disease. Clin Exp Allergy 31(6):893–902

Fabbri LM, Romagnoli M, Corbetta L, Casoni G, Busljetic K, Turato G, Ligabue G, Ciaccia A, Saetta M, Papi A (2003) Differences in airway inflammation in patients with fixed airflow obstruction due to asthma or chronic obstructive pulmonary disease. Am J Respir Crit Care Med 167(3):418–424

GOLD (2014) Global Initiative for Chronic Obstructive Lung Disease: global strategy for the diagnosis, management and prevention of chronic obstructive pulmonary disease. http://www.gold copd.com. Accessed 20 Oct 2017

Górska K, Krenke R, Korczyński P, Domagała-Kulawik J, Nejman-Gryz P, Kosciuch J, Hildebrand K, Chazan R (2008) Eosinophilic airway inflammation in chronic obstructive pulmonary disease and asthma. J Physiol Pharmacol 59(6):261–270

Hattotuwa KL, Gizycki MJ, Ansari TW, Jeffery PK, Barnes NC (2002) The effects of inhaled fluticasone on airway inflammation in chronic obstructive pulmonary disease: a double-blind, placebo-controlled biopsy study. Am J Respir Crit Care Med 165 (12):1592–1596

Hogg JC (2004) Pathophysiology of airflow limitation in chronic obstructive pulmonary disease. Lancet 364:709–721

Hurst JR, Veestbo J, Anzueto A, Müllerova H, Tal-Singer R, Miller B, Lomas DA, Agusti A, MacNee W, Calverley P, Rennard S, Wouters EFM, Wedzicha JA (2010) Susceptibility to exacerbation in chronic obstructive pulmonary disease. N Engl J Med 363(12):1128–1138

Jeffery PK (2001) Remodeling in asthma and chronic obstructive lung disease. Am J Respir Crit Care Med 164(10 Pt2):S28–S38

Lams BEA, Sousa AR, Rees PJ, Lee TH (1998) Immunopathology of the small-airway submucosa in smokers with and without chronic obstructive pulmonary disease. Am J Respir Crit Care Med 158(5):1518–1523

Lams BE, Sousa AR, Rees PJ, Lee TH (2000) Subepithelial immunopathology of the large airways in smokers with and without chronic obstructive pulmonary disease. Eur Respir J 15(3):512–516

Lieskier JW, Hacken NH, Zeinstra-Smith M, Rutgers SR, Postma DS, Timens W (2009) Reticular basement membrane in asthma and COPD: similar thickness, yet different composition. Int J Chron Obstruct Pulmon Dis 4:127–135

Lopez AD, Shibuya K, Rao C, Mathers CD, Hansell AL, Held LS, Schmid V, Buist S (2006) Chronic obstructive pulmonary disease: current burden and future projections. Eur Respir J 27(2):397–412

Mathers CD, Loncor D (2006) Projections of global mortality and burden of disease from 2002 to 2030. PLoS Med 3(11):e442

O'Donnell R, Breen D, Wilson S, Djukanowic R (2006) Inflammatory cells in the airways in COPD. Thorax 61 (5):448–454

O'Shaughnessy T, Ansari TW, Barnes NC, Jeffery PK (1997) Inflammation in bronchial biopsies of subjects with chronic bronchitis: inverse relationship of CD8+ T lymphocytes with FEV1. Am J Respir Crit Care Med 155(3):852–857

Olloquequi J, Ferrer J, Montes J, Rodríguez E, Montero MA, García-Valero J (2010) Differential lymphocyte infiltration in small airways and lung parenchyma in COPD patients. Respir Med 104(9):1310–1318

Petrescu F, Biciuşcă V, Octavia IP, Forţofoiu CM (2013) Cellular basis of bronchial inflammation in chronic obstructive pulmonary disease. J Cur Health Sci 39 (2):61–67

Saetta M, Baraldo S, Corbino L, Turato G, Braccioni F, Rea F, Cavallesco G, Tropeano G, Mapp CE, Maestrelli P, Ciaccia A (1999) CD8+ cells in the lungs of smokers with chronic obstructive pulmonary disease. Am J Respir Crit Care Med 160(2):711–717

Sohal SS, Reid D, Soltani A, Ward C, Weston S, Muller HK, Wood-Baker R, Walters EH (2010) Reticular basement membrane fragmentation and potential epithelial mesenchymal transition is exaggerated in the airways of smokers with chronic obstructive pulmonary disease. Respirology 15(6):930–938

Sohal SS, Reid D, Soltani A, Ward C, Weston S, Muller HK, Wood-Baker R, Walters EH (2011) Evaluation of epithelial mesenchymal transition in patients with chronic obstructive pulmonary disease. Respir Res 12(1):130–137

Sohal SS, Soltani A, Reid D, Ward C, Wills K, Muller HK, Walters EH (2014) A randomized controlled trial of inhaled corticosteroids (ICS) on markers of epithelial – mesenchymal transition (EMT) in large airways samples in COPD: an exploratory proof on concept study. Int J Chron Obstruct Pulmon Dis 9:533–542

Soltani A, Reid DW, Sohal SS, Wood-Baker R, Weston S, Muller HK, Walters EH (2010) Basement membrane and vascular remodeling in smokers and chronic obstructive pulmonary disease: a cross-sectional study. Respir Res 11(1):105

Soltani A, Wood-Baker R, Sukhwinder S, Sohal H, Muller K, Reid D, Walters EH (2012) Reticular basement membrane vessels are increased in COPD bronchial mucosa by both factor VIII and collagen IV immunostaining and are hyperpermeable. J Allergy (Cairo) 2012:958383

Stanescu D, Senna A, Veriter C, Kostianev S, Calcagni PG, Fabbri LM, Maestrelli P (1996) Airways obstruction, chronic expectoration and rapid decline of FEV1 in smokers are associated with increased levels of sputum neutrophils. Thorax 51(3):267–271

Turato G, Di Stefano A, Maestrelli P, Mapp CE, Ruggieri MP, Rogerri A, Fabbri LM, Saetta M (1995) Effect of smoking cessation on airway inflammation in chronic bronchitis. Am J Respir Crit Care Med 152 (4):1262–1267

Advs Exp. Medicine, Biology - Neuroscience and Respiration (2018) 38: 53–60
DOI 10.1007/5584_2017_109

Published online: 22 September 2017

Association Between Vitamin D and Carboxy-Terminal Cross-Linked Telopeptide of Type I Collagen in Children During Growth Hormone Replacement Therapy

Ewelina Witkowska-Sędek, Anna Stelmaszczyk-Emmel, Anna Kucharska, Urszula Demkow, and Beata Pyrżak

Abstract

Growth hormone and insulin-like growth factor-1 (IGF-1) play a crucial role in the regulation of bone turnover. Adequate vitamin D status supports proper bone remodeling, leading to normal longitudinal bone growth and normal peak bone mass. The aim of this study was to evaluate the association between serum 25-hydroxyvitamin D [25(OH)D] and carboxy-terminal cross-linked telopeptide of type I collagen (ICTP) in children and adolescents with growth hormone deficiency at baseline and during recombinant human growth hormone (rhGH) replacement therapy. The study was prospective and included 30 children and adolescents aged 5 to 17 years. Concentrations of 25(OH)D, ICTP, and IGF-1 were measured at baseline and during the first year of rhGH therapy. Baseline serum 25(OH)D concentration correlated with ICTP concentrations during the first trimester of rhGH therapy ($r = 0.38$, $p < 0.050$); the correlation was stronger in the second trimester of therapy ($r = 0.6$, $p = 0.002$). We conclude that proper vitamin D status is important in reaching the adequate dynamics of bone remodeling during growth, which is essential to achieve a catch-up growth during rhGH therapy.

Keywords

Bone turnover · Children · Collagen · Growth hormone deficiency · Growth hormone replacement therapy · Vitamin D

E. Witkowska-Sędek (✉), A. Kucharska, and B. Pyrżak
Department of Paediatrics and Endocrinology, Medical University of Warsaw, 63A Żwirki and Wigury Street, 02-091 Warsaw, Poland
e-mail: ewelina.witkowska-sedek@wum.edu.pl

A. Stelmaszczyk-Emmel and U. Demkow
Department of Laboratory Diagnostics and Clinical Immunology of Developmental Age, Medical University of Warsaw, Warsaw, Poland

1 Introduction

Bone turnover is regulated by a variety of factors that act systemically or locally. Some of them are nutritional factors, including 25-hydroxyvitamin D [25(OH)D] (Schwetz et al. 2017; Bonjour et al. 2014). Hypovitaminosis D is common among children and adolescents, especially those who do not receive supplementation during autumn and winter seasons. The relationship between bone turnover markers and 25(OH)D status is not quite clear, but in recent years it has been studied extensively in children, adolescents, and adults (Larijani et al. 2016; Thiering et al. 2015; Napoli et al. 2014; Fares et al. 2003). Bone turnover markers are widely used in medicine. They are a useful tool in evaluating bone health, diagnosing, and in treatment monitoring in primary and secondary bone diseases in oncology, rheumatology, nephrology, cardiology, and endocrinology (Shetty et al. 2016; Fardellone et al. 2014; Wheater et al. 2013; Lijnen et al. 2012; Coleman et al. 2011). In pediatric endocrinology, bone turnover markers are used in the evaluation of skeletal metabolism and growth (Bowden et al. 2016; Szulc et al. 2000). These markers could be particularly helpful in metabolic disorders related to diabetes, obesity, and in those patients who are treated with corticosteroids (Matusik et al. 2015; Zhukouskaya et al. 2014). In children with growth hormone deficiency, selected bone turnover markers could be used to monitor the efficacy of recombinant human growth hormone (rhGH) therapy (Andersson et al. 2015; Kandemir et al. 2002). The interpretation of bone turnover marker values could be difficult due to their variability depending on age, pubertal stage, current growth velocity, mineral accrual, nutritional status, and the influence of hormones, especially growth hormone/insulin-like growth factors (GH/IGFs) axis and sex steroids (Locatelli and Bianchi 2014; Federico et al. 2003; Szulc et al. 2000).

Classical biochemical markers of bone turnover are divided into two groups: bone formation markers and bone resorption markers. Bone formation markers are products of osteoblasts' activity during different phases of their development. These markers are measured in serum or plasma and they are categorized as by-products of collagen synthesis [carboxy- (PICP) and amino-terminal (PINP) propeptides of procollagen type I], osteoblast enzymes [total alkaline phosphatase (ALP) and bone-specific alkaline phosphatase (BALP)], and matrix proteins [osteocalcin (OC)]. Bone resorption markers, on the other side, arise during the resorption phase of bone remodeling. They can be measured in serum or urine. They are mainly products of collagen degradation by osteoclasts, such as telopeptides of type I collagen (carboxy-terminal: CTX-1 and ICTP, also referred to as CTX-matrix metalloproteinases - MMP; and amino-terminal: NTX-1), hydroxyproline and pyridinium crosslinks [pyridinoline (PYD), deoxypyridinoline (DPD)]. Bone resorption markers are also non-collagenous proteins (bone sialoprotein) and osteoclastic enzymes [tartrate-resistant acid phosphatase – isoform 5b (TRAP5b), cathepsin K] (Chapurlat and Confavreux 2016; Wheater et al. 2013; Szulc et al. 2000). According to Shetty et al. (2016), the group of bone resorption markers also includes osteocyte activity markers such as a receptor activator of nuclear factor kappa-B ligand (RANKL), osteoprotegerin (OPG), Dickkopf-related protein-1 (DKK-1), and sclerostin. Recently, some authors have described potential new bone markers such as markers of bone metabolism (periostin and sphingosine-1-phosphate), hormones (fibroblast growth factor-23, klotho, osteocalcin), and micro-RNAs (miR-148a, miR125b) (Chapurlat and Confavreux 2016; Garnero 2014).

The aim of the present study was to evaluate the association between serum 25(OH)D and carboxy-terminal cross-linked telopeptide of type I collagen (ICTP) levels in children with growth hormone deficiency at baseline and during the first year of rhGH therapy. We also sought to determine if ICTP could help predict the growth response to rhGH therapy in children and adolescents with growth hormone deficiency.

2 Methods

The study was approved by the Bioethics Committee of Medical University of Warsaw, Poland. All patients or their parents gave written informed consent for participation in the study. The study was conducted at the Department of Paediatrics and Endocrinology of the Medical University of Warsaw in Poland in the years 2013–2016 and it included 30 children and adolescents (9 girls and 21 boys) aged 4.8–16.6 years (mean 11.4 ± 3.2 years) with growth hormone deficiency. The study was prospective in nature as it evaluated the first year of rhGH treatment. Growth hormone deficiency was diagnosed as growth hormone concentrations below 10 ng/ml in the following three tests: one test evaluating spontaneous growth hormone secretion at night during sleep and two stimulation tests with different stimuli such as clonidine, insulin, glucagon, or arginine. The maximum growth hormone secretion in a particular patient was defined as the highest concentration of growth hormone in any of the three tests. Growth rate at baseline was calculated based on the measurement of height taken twice at six-month intervals. Anthropometric measurements were taken at baseline and after 3, 6, and 12 months of rhGH treatment. Based on those measurements, growth rate in the first year of rhGH treatment was calculated. Bone age was evaluated in each patient using the Greulich and Pyle method at baseline and after 12 months of rhGH treatment (Greulich and Pyle 1959). 25 (OH)D concentration was measured at baseline and after 6 and 12 months of rhGH treatment. ICTP and insulin-like growth factor-1 (IGF-1) were measured at baseline and after 3, 6, and 12 months of rhGH treatment. The mean dose of rhGH in the first 12 months of therapy was 0.187 mg/kg/week. Apart from rhGH treatment, all patients received cholecalciferol in doses of 1,000 IU/day within the first 12 months of rhGH therapy.

2.1 Biochemical Evaluation

GH and IGF-1 concentrations were measured by immunoassay using Immulite Analyzer from Siemens (Erlangen, Germany); 25(OH)D concentration was measured by immunoassay using Architect Analyzer (Abbott Diagnostics; Abbott Park, IL); and ICTP concentration was measured by Human ICTP ELISA kit using Asys UVM 340 Microplate Reader from Biochrom (Cambridge, England).

2.2 Statistical Elaboration

Data are reported as means ±SD, median and interquartile ranges, or as percentages. Data normality was checked by the Shapiro-Wilk normality test. Comparisons between baseline and treatment values were conducted using the Wilcoxon signed-rank test for non-parametric parameters and a *t*-test for parametric parameters. In addition, Spearman and Pearson's correlation analyses were performed. A p-value <0.05 was used to indicate statistically significant differences. Statistical elaboration was conducted with a commercial Statistica v13.1 package (StatSoft; Tulsa, OK).

3 Results

The maximum growth hormone secretion in the study group was 7.4 ± 1.7 ng/ml. In the first year of rhGH therapy, the mean growth rate increased significantly from 4.6 ± 1.3 cm/year at baseline to 9.3 ± 1.8 cm/year ($p < 0.00001$). Height deficit declined during that time from 2.5 ± 0.4 cm at baseline to −1.9 ± 0.5 cm ($p < 0.00001$). A reduction in height deficit was significant as early as after the first trimester of rhGH therapy ($p < 0.00002$). The mean bone age increased significantly after the start of rhGH therapy from 8.9 ± 3.1 years at baseline to 10.6 ± 3.0 years ($p < 0.000001$). The characteristics of the study group at baseline and during the first 12 months rhGH therapy are shown in Table 1. The mean 25(OH)D concentration was 24.2 ± 7.8 ng/ml at baseline and increased significantly during the first 6 months of rhGH therapy to 28.5 ± 5.3 ng/ml ($p = 0.014$). 25(OH)D concentration after six and 12 months of rhGH therapy was not appreciably different. The mean ICTP value at baseline was 0.72 ± 0.42 ng/ml and decreased after the first 3 months of rhGH therapy to 0.68 ± 0.27 ng/ml; the difference was insignificant. After 6 months of rhGH therapy, ICTP concentration increased to 0.79 ± 0.49 ng/ml, which was higher than the baseline value ($p = 0.02$). After 12 months of rhGH therapy, ICTP concentration slightly decreased to 0.74 ± 0.11 ng/ml; the figure was still significantly greater than the baseline level ($p < 0.001$), but not so compared with the level present 6 months after rhGH therapy. As expected, IGF-1 concentrations increased significantly in the first 12 months of rhGH treatment ($p < 0.000001$), and the difference was significant as early as after 3 months of therapy ($p = 0.000002$). Table 2 details the biochemical indices evaluated at baseline and during rhGH treatment.

The serum 25(OH)D concentration did not correlate with that of ICTP at baseline, but it did correlate with ICTP changes during the first 3 months of rhGH therapy ($r = 0.38$, $p < 0.05$, Fig. 1). Likewise, there was no correlation between the serum 25(OH)D and ICTP after 6 months of rhGH therapy, but 25(OH)D did correlate with ICTP changes between the three and 6 months of rhGH therapy ($r = 0.6$, $p = 0.002$) (Fig. 2). The latter correlation was no longer present after 12 months of therapy. In addition, we found that ICTP changes between baseline and first 6 months of fhGH therapy correlated positively with growth rate (cm/year) ($r = 0.43$, $p = 0.042$).

4 Discussion

Growth hormone, insulin-like growth factors, and IGF-binding proteins are essential proteins, among many other factors regulating bone turnover, for achieving a normal longitudinal bone growth and peak bone mass (Schwetz et al. 2017; Bonjour et al. 2014). During puberty GH/IGF axis's effects on bone are supported by sex steroids action (Locatelli and Bianchi 2014; Mrak et al. 2007). Bone development is also

Table 1 Characteristics of the study group at baseline and during recombinant human growth hormone (rhGH) replacement therapy

Parameter	Baseline	Three months	Six months	Twelve months
Height SDS for age	−2.51 ± 0.42	−2.30 ± 0.41	−2.16 ± 0.37	−1.89 ± 0.49
Body mass SDS for height-age	0.12 ± 0.77	0.01 ± 0.80	−0.08 ± 0.76	0.02 ± 0.65
BMI SDS for height-age	0.18 ± 1.09	0.12 ± 1.04	−0.06 ± 0.98	0.07 ± 0.84
Bone age (years)	8.92 ± 3.07			10.58 ± 3.02
Growth rate (cm/year)	4.62 ± 1.29			9.34 ± 1.80

Data are means ±SD. *BMI* body mass index (kg/m^2), *SDS* standard deviation score

Table 2 Biochemical indices at baseline and during recombinant human growth hormone (rhGH) replacement therapy

Parameters	Baseline	Three months	Six months	Twelve months
25(OH)D (ng/ml)	24.2 ± 7.8	–	28.5 ± 5.3	28.5 ± 7.7
ICTP (ng/ml)	0.72 ± 0.42	0.68 ± 0.27	0.79 ± 0.49	0.74 ± 0.11
IGF-1 (ng/ml)	167.4 ± 93.1	347.0 ± 192.2	375.9 ± 210.3	453.2 ± 242.6

Data are means ±SD. *25(OH)D* 25-hydroxyvitamin D, *ICTP* carboxy-terminal cross-linked telopeptide of type I collagen, *IGF-1* insulin-like growth factor-1

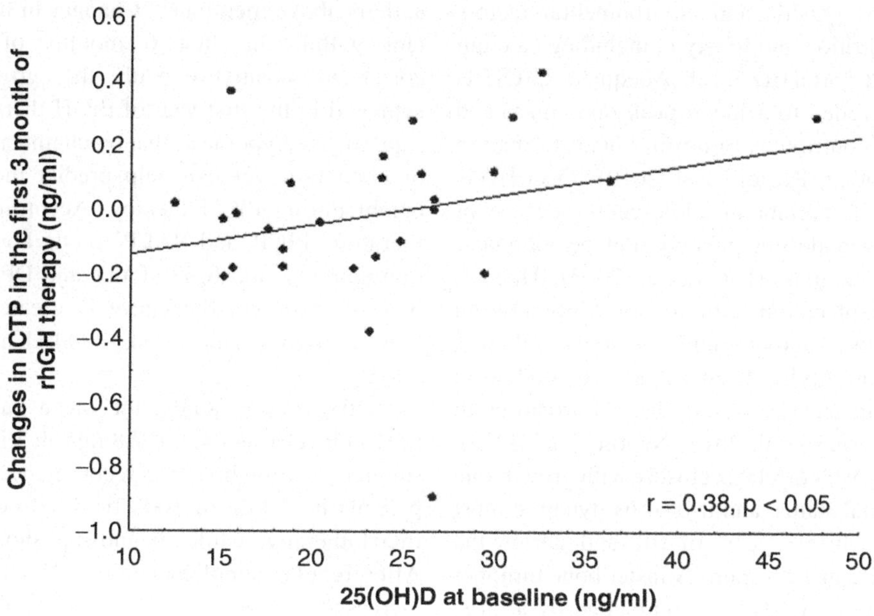

Fig. 1 Correlation between the baseline 25(OH)D level and changes in the concentration of carboxy-terminal cross-linked telopeptide of type I collagen (ICTP) during the first 3 months of recombinant human growth hormone (rhGH) replacement therapy

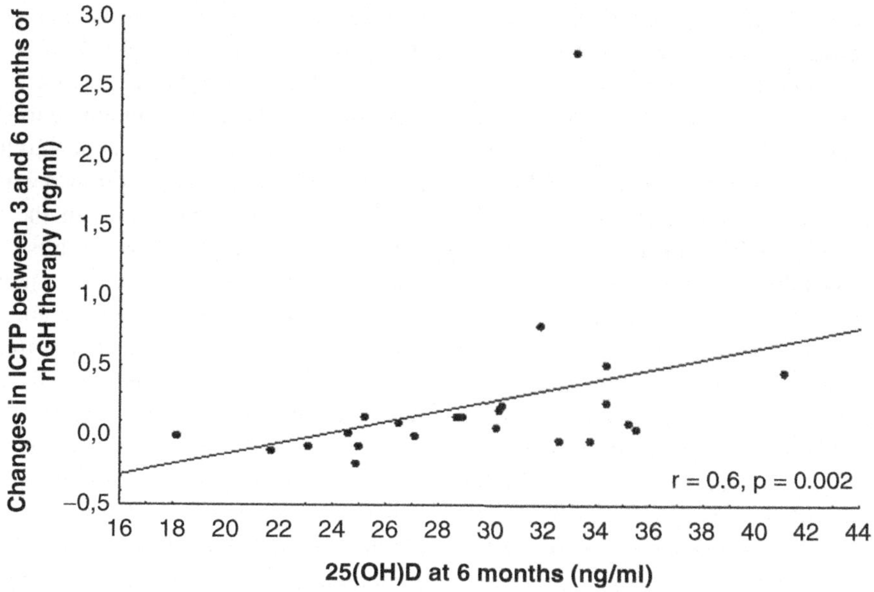

Fig. 2 Correlation between the 25(OH)D level at 6 months and changes in the concentration of carboxy-terminal cross-linked telopeptide of type I collagen (ICTP)between three and 6 months of rhGH therapy

affected by genetic and environmental factors such as nutrition and lifestyle, including calcium intake and 25(OH)D level. Adequate 25(OH)D status is needed to achieve peak bone mass and 25(OH)D contributes to normal bone formation and resorption (Rizzoli et al. 2010). 25(OH)D is especially important in adolescents because of rapid growth during puberty and development of peak bone mass (Fares et al. 2003). There is no clear explanation of the relationship between bone turnover markers and 25(OH)D, although this relationship has been extensively studied in recent years (Schwetz et al. 2017, Larijani et al. 2016, Thiering et al. 2015, Napoli et al. 2014). Bone turnover correlates closely with growth rate and pubertal status, and it is at its dynamic most during the first 2 years of life and during the adolescence growth spurt. A faster bone turnover in these two periods is accompanied by a large increase in the concentration of bone remodeling markers (Fares et al. 2003; Federico et al. 2003). A study by van Coeverden et al. (2002) shows that markers of bone formation (BALP, OC, and PICP) and markers of bone resorption (ICTP) in children and adolescents reach the maximum values at pubertal stage G4 in boys and Th 3 in girls. The markers content parallels the increases in sex steroids and IGF-1 concentrations.

In children with growth hormone deficiency, content of bone turnover markers declines as a consequence of decreased bone metabolism, which leads to decreased bone mineral density (Szulc et al. 2000; Vihervuori et al. 1997; Kubo et al. 1995). GH therapy stimulates the secretion of insulin-like growth factors-1 and IGF-binding protein-3, which activates bone turnover. After onset of growth hormone therapy, markers of both bone formation and resorption increase with a peak between the three and 6 months of treatment and then remain elevated over the following months of therapy (Andersson et al. 2015; Mrak et al. 2007; Kandemir et al. 2002; Baroncelli et al. 2000).

In the present study, ICTP values increased significantly after onset of rhGH therapy, with a peak at 6 months of therapy and remained still higher than the baseline value at 12 months of therapy. This is in line with the studies of other authors above mentioned. Changes in ICTP content within the first 6 months of therapy correlated positively with the growth rate achieved in the first year of rhGH therapy. That supports the hypothesis that biochemical markers of bone turnover can help predict the gain in height during rhGH therapy. According to the literature, PICP, and BALP among bone formation markers, and ICTP, PYD, and DPD among bone resorption markers have the greatest predictive strength (Szulc et al. 2000; Kubo et al. 1995).

In the present study, the mean baseline 25 (OH)D level was 24.2 ± 7.8 ng/ml. Vitamin D3 supplementation had been recommended to most patients in a dose of 1000 IU/d before onset of rhGH therapy, while diagnosing short stature. After regular supplementation, 25(OH)D level increased to 28.5 ± 5.3 ng/ml at 6 months, reaching a plateau for the next months. The baseline 25(OH)D values were not associated with the baseline ICTP values. We failed to find a direct correlation between the 25(OH)D and ICTP levels in any of the measurements at any evaluated time point, but we found a correlation between the 25(OH)D levels and changes in ICTP concentrations during rhGH therapy. Baseline 25(OH)D correlated positively with an increase in ICTP concentrations during the first 3 months of rhGH therapy, while 25(OH)D level after 6 months of therapy correlated with changes in ICTP concentrations between the three and 6 months of therapy. Those results are in line with some recent studies that have shown a positive correlation between the serum 25(OH)D level and bone turnover. Such a correlation has been found in healthy children and adolescents during rapid growth during puberty, and in children with GHD during catch-up growth after onset of rhGH therapy (Larijani et al. 2016; Thiering et al. 2015; Fares et al. 2003). These results confirm that proper 25(OH)D status is important in the bone remodeling during growth and helps improve the growth rate. Vitamin D3 supplementation is important during rapid growth in healthy children and adolescents and in children treated with rhGH.

5 Conclusions

Bone resorption increases significantly after onset of rhGH therapy. Early increases in the concentration of carboxy-terminal cross-linked telopeptide of type I collagen correlate with the growth rate during rhGH therapy. Correct vitamin D3 status contributes to an increase in bone turnover after onset of rhGH therapy, which is necessary to achieve catch-up growth. Vitamin D3 supplementation should be recommended to patients with growth hormone deficiency treated with rhGH.

Conflicts of Interest The authors declare no conflicts of interest in relation to this article.

References

Andersson B, Swolin-Eide D, Magnusson P, Albertsson-Wikland K (2015) Short-term changes in bone formation markers following growth hormone (GH) treatment in short prepubertal children with a broad range of GH secretion. Clin Endocrinol 82 (1):91–99

Baroncelli GI, Bertelloni S, Ceccarelli C, Cupelli D, Saggese G (2000) Dynamics of bone turnover in children with growth hormone (GH) deficiency treated with GH until final height. Eur J Endocrinol 142:549–556

Bonjour JP, Kohrt W, Levasseur R, Warren M, Whiting S, Kraenzlin M (2014) Biochemical markers for assessment of calcium economy and bone metabolism: application in clinical trials from pharmaceutical agents to nutritional products. Nutr Res Rev 27:252–267

Bowden SA, Akusoba CI, Hayes JR, Mahan JD (2016) Biochemical markers of bone turnover in children with clinical bone fragility. J Pediatr Endocrinol Metab 29(6):715–722

Chapurlat RD, Confavreux CB (2016) Novel biological markers of bone: from bone metabolism to bone physiology. Rheumatology 55:1714–1725

Coleman R, Costa L, Saad F, Cook R, Hadji P, Terpos E, Garnero P, Brown J, Body JJ, Smith M, Lee KA, Major P, Dimopoulos M, Lipton A (2011) Consensus on the utility of bone markers in the malignant bone disease setting. Crit Rev Oncol Hematol 80:411–432

Fardellone P, Séjourné A, Paccou J, Goëb V (2014) Bone remodeling markers in rheumatoid arthritis. Mediat Inflamm 2014:484280. https://doi.org/10.1155/2014/484280

Fares JE, Choucair M, Nabulsi M, Salamoun M, Shahine CH, Fuleihan GEH (2003) Effect of gender, puberty, and vit. D status on biochemical markers of bone remodeling. Bone 33:242–247

Federico G, Baroncelli GI, Vanacore T, Fiore L, Saggese G (2003) Pubertal changes in biochemical markers of growth. Horm Res 60(Suppl 1):46–51

Garnero P (2014) New developments in biological markers of bone metabolism in osteoporosis. Bone 66:46–55

Greulich WW, Pyle SI (1959) Radiographic atlas of skeletal development of the hand and wrist, 2nd edn. Stanford University Press, Stanford

Kandemir N, Gonc EN, Yordam N (2002) Responses of bone turnover markers and bone mineral density to growth hormone therapy in children with isolated growth hormone deficiency and multiple pituitary hormone deficiencies. J Pediatr Endocrinol Metab 15 (6):809–816

Kubo T, Tanaka H, Inoue M, Kanzaki S, Seino Y (1995) Serum levels of carboxyterminal propeptide of type I procollagen and pyridinoline crosslinked telopeptide of type I collagen in normal children and children with growth hormone (GH) deficiency during GH therapy. Bone 17:397–401

Larijani B, Hossein-Nezhad A, Feizabad E, Maghbooli Z, Adibi H, Ramezani M, Taheri E (2016) Vitamin D deficiency, bone turnover markers and causative factors among adolescents: a cross-sectional study. J Diabetes Metab Disord 15:46. https://doi.org/10.1186/s40200-016-0266-2

Lijnen PJ, Maharani T, Finahari N, Prihadi JS (2012) Serum collagen markers and heart failure. Cardiovasc Hematol Disord Drug Targets 12(1):51–55

Locatelli V, Bianchi VE (2014) Effect of GH/IGF-1 on bone metabolism and osteoporosis. Int J Endocrinol 2014:235060. https://doi.org/10.1155/2014/235060

Matusik P, Olszanecka-Glinianowicz M, Chudek J, Małecka-Tendera E (2015) Bone turnover markers in the obese children – relations to gender, body composition and leptin level. Pediatr Endocrinol Diabetes Metab 21(4):154–161

Mrak E, Villa I, Lanzi R, Losa M, Guidobono F, Rubinacci A (2007) Growth hormone stimulates osteoprotegerin expression and secretion in human osteoblast-like cells. J Endocrinol 192:639–645

Napoli N, Strollo R, Sprini D, Maddaloni E, Rini GB, Carmina E (2014) Serum 25-OH vit. D in relations to bone mineral density and bone turnover. Int J Endocrinol 2014:487463. https://doi.org/10.1155/2014/487463

Rizzoli R, Bianchi ML, Garabédian M, McKay HA, Moreno LA (2010) Maximizing bone mineral mass gain during growth for the prevention of fractures in the adolescents and the elderly. Bone 46:294–305

Schwetz V, Trummer C, Pandis M, Grübler MR, Verheyen N, Gaksch M, Zittermann A, März W, Aberer F, Lang A, Treiber G, Friedl C, Obermayer-Pietsch B, Pieber TR, Tomaschitz A, Pilz S (2017)

Effects of vit. D supplementation on bone turnover markers: a randomized controlled trial. Forum Nutr 9 (5). https://doi.org/10.3390/nu9050432

Shetty S, Kapoor N, Bondu JD, Thomas N, Paul TV (2016) Bone turnover markers: emerging tool in the management of osteoporosis. Indian J Endocrinol Metab 20(6):846–852

Szulc P, Seeman E, Delmas PD (2000) Biochemical measurements of bone turnover in children and adolescents. Osteoporosis Int 11(4):281–294

Thiering E, Brüske I, Kratzsch J, Hofbauer LC, Berdel D, von Berg A, Lehmann I, Hoffmann B, Bauer CP, Koletzko S, Heinrich J (2015) Associations between serum 25-hydroxyvit. D and bone turnover markers in a population based sample of German children. Sci Rep 5:18138. https://doi.org/10.1038/srep18138

van Coeverden SC, Netelenbos JC, de Ridder CM, Roos JC, Popp-Snijders C, Delemarre-van de Waal HA (2002) Bone metabolism markers and bone mass in healthy pubertal boys and girls. Clin Endocrinol 57:107–116

Vihervuori E, Turpeinen M, Siimes MA, Koistinen H, Sorva R (1997) Collagen formation and degradation increase during growth hormone therapy in children. Bone 20:133–138

Wheater G, Elshahaly M, Tuck SP, Datta HK, van Laar JM (2013) The clinical utility of bone marker measurements in osteoporosis. J Transl Med 11:201. https://doi.org/10.1186/1479-5876-11-201

Zhukouskaya VV, Eller-Vainicher C, Shepelkevich AP, Dydyshko Y, Cairoli E, Chiodini I (2014) Bone health in type 1 diabetes: focus on evaluation and treatment in clinical practice. Adv Endocrinol Invest 38 (9):941–950

Advs Exp. Medicine, Biology - Neuroscience and Respiration (2018) 38: 61–69
DOI 10.1007/5584_2017_124

Published online: 19 November 2017

Weight Disorders in Short Children

Paweł Tomaszewski, Anna Majcher, Katarzyna Milde, and Romuald Stupnicki

Abstract

This study aimed to determine the degree of adiposity and prevalence of weight disorders in short children and to assess relationships between selected birth parameters and actual body fat content and body mass indicators. The study encompassed 85 girls and 97 boys, aged 7–14 years, with height below the 10th centile of the age and sex standard. In addition, 609 normal-height children (height between 25th and 75th centile) served as a comparison group. The children underwent measurements of body mass and height, skinfold thickness, and waist circumference. Detailed history on pregnancy and neonatal status of short children was collected. Body mass index, waist-to-height ratio, and body fat percentage (%FAT) were assessed. We found that about 27% of short children had insufficient %FAT and about 14% of them had excessive %FAT, regardless of sex. In terms of BMI, 24% of short children were underweight and only 6% were overweight or obese, while in normal-height children these percentages were 6 and 27.5%, respectively. Abdominal obesity was observed in about 5% of short children and 18% of normal-height children ($p < 0.001$). There was no association between birth parameters and somatic indicators in short children. We conclude that the prevalence of overweight and obesity, including abdominal obesity, among short children is much lower than that in normal-height children. The insufficient fat tissue and body mass, observed in one out of four short children, are particularly disturbing in the context of ensuring their correct development.

Keywords

Abdominal obesity · Adipose tissue · Birth parameters · Body fat content · Body mass index · Central obesity · Short children · Weight disorder

P. Tomaszewski (✉) and K. Milde
Department of Biometry, The Józef Piłsudski University of Physical Education in Warsaw, Warsaw, Poland
e-mail: pawel.tomaszewski@awf.edu.pl

A. Majcher
Department of Pediatrics and Endocrinology, Independent Public Children's Clinical Hospital, Warsaw Medical University, Warsaw, Poland

R. Stupnicki
The Halina Konopacka Higher School of Physical Culture and Tourism in Pruszków, Pruszków, Poland

1 Introduction

Depending on the criterion, between 3 and 10% of the population are abnormally short. Growth disorders and short stature are among the most frequent reasons for consultations with a pediatric endocrinologist (Wudy et al. 2005). While studies most often indicate genetic factors as the cause of short stature (Pasquino et al. 2001; Bierich 1992), it is difficult to explain the etiology of the disorder in young patients. Research indicates that only 5% of students with body height below the third centile have the previously unrecognized disorders that are responsible for short stature (Voss et al. 1992). Consequently, short children are usually diagnosed with a constitutional delay of growth and puberty, and familial or idiopathic short stature. Another cause of short stature during childhood and adolescence, and a short final body height is low birth weight. Greco et al. (1995) have estimated that the risk of short stature at the age of 7 years is over 2.5 times higher in children with low birth weight compared to children with normal birth weight. Furthermore, Wudy et al. (2005) have shown that short stature during adolescence correlates with low fat mass and low body mass index (BMI). This creates a specific contradiction, as other studies indicate that low birth weight is associated with increased risk of obesity and its metabolic and circulatory comorbidities (Monasta et al. 2010; Rugholm et al. 2005).

Prevalence of obesity and overweight in the general population is well documented, but few studies have addressed the adiposity and the prevalence of weight disorders among short children. Such studies are often based on the relatively small samples of patients (Tomaszewski et al. 2017) and a rather large age range (Thibault et al. 1993), or on insufficiently strict (over the 10th centile) cut-off points for short stature (Freedman et al. 2004a, b). The effect of low birth weight on the future development of a child is known, but the existing studies have not investigated the association between birth parameters and adiposity or the weight-height proportions in short children (Monasta et al. 2010; Rugholm et al. 2005; Greco et al. 1995). It seems that such data could provide valuable information on the causes of weight disorders among short children. Therefore, the aim of the present study was to determine adiposity and the prevalence of weight disorders among children with short stature and to assess the association between selected birth parameters and the actual body fat content and body mass indicators in short children.

2 Methods

The study was conducted in accordance with the guidelines provided by the Declaration of Helsinki for Human Research and was accepted by the institutional Ethics Committee.

2.1 Participants

The study was conducted in five randomly selected state schools in Warsaw, Poland. The children and their parents were informed about the aim and course of the study. A permission from both children and their legal caretakers was required for inclusion in the study. There were 182 children, aged 7–14 years (85 girls and 97 boys), enrolled into the study. The children had height insufficiency, defined as height below the 10th centile for the appropriate age and sex. They were not on hormonal therapy. Additionally, 609 age-matched, normal-height children (height between 25th and 75th centile; 306 girls and 303 boys) were included for comparison. The cut-off points for height were selected according to the Polish reference values for somatic traits, which have been established based on a sample of about 35,000 children (Dobosz 2012).

2.2 Study Protocol

Participants underwent measurements of selected somatic traits, which were performed by the trained personnel using standard methodology and equipment (Hall et al. 2007). In addition, data concerning pregnancy length, birth weight, and birth length of short children were collected based on the medical records and declarations of the children' parents or caretakers. Height was measured using anthropometry (Holtain; Crymych, UK) with accuracy to 0.1 cm. Body mass was measured using a Tanita TBF-300A (Tokyo, Japan) body composition analyzer with accuracy to 0.1 kg. The thickness of skinfolds on the triceps, under the scapula, and over the ilium wing was measured using a skinfold caliper with accuracy to 0.5 mm. Three measurements were taken on each site and the final result was averaged. Finally, waist circumference was measured at the navel during a short apnea using an anthropometric tape. The parameters measured were used to calculate BMI and waist-to-height ratio (WHtR) and to estimate %FAT according to Slaughter et al.'s (1988) equations. Since complete medical records were unavailable for some short children, and their parents or caretakers provided insufficient information, the evaluation of birth weight and birth length was ultimately performed for 47 (55%) girls and 46 (47%) boys with short stature. Data on the pregnancy length were available for 40 short girls (47%) and 38 (39%) short boys. Table 1 presents the detailed data concerning the assessed somatic traits and birth parameters in short and normal-height children.

Participants were classified according to adiposity and weight-to-height proportions based on the reference values for the Polish population and the suggested cut-off points for %FAT and BMI (Tomaszewski et al. 2013). The classification was performed using polynomial equations as functions of age computed for boys and girls, and BMI cut-off points corresponding to the WHO recommended values, i.e., BMI <18.5 – underweight, BMI >25.0 – overweight, and BMI >30.0 – obesity. On this basis, children with decreased, normal, or increased adipose tissue and obesity were distinguished, and also children who, depending on their BMI, were characterized by underweight, normal weight, overweight, or obesity (for details see Tomaszewski et al. 2013). The WHtR $\geq$ 0.5 criterion (Browning et al. 2010), and those taking into account the age, sex, and adiposity (Stupnicki et al. 2013) was applied to assess the prevalence of abdominal adiposity. In the latter method, children with the excessive body fat content were classified according to reference WtHR values defined as means ±2SD. The WtHR value within normal limits was considered to indicate a relatively uniform, peripheral fat distribution, while values above the upper limit indicated central (abdominal) fat deposition. These limits are 0.480 and 0.460 for boys, and 0.495 and 0.455 for girls, for ages 7 and 14, respectively. Prevalence of different categories of BMI also was assessed for short children with birth weight or birth length below the fifth centile (2600 g and 46.8 cm for boys and 2500 g and 46.1 cm for girls) of the WHO (2006) standards and for the children born before the 37th week.

2.3 Statistical Elaboration

Data are presented as means ±SD and percentages ±SE. Differences in the prevalence of each category of %FAT, BMI, or WHtR among short girls and boys and between short and normal-height children were assessed using the chi-square test; Cramer's V was used as a measure of effect size. Correlations between pregnancy length, birth weight, birth length, and the normalized values (*z*-scores) of %FAT and BMI and the values of WHtR were assessed in short children using Pearson's correlation coefficient. Statistical significance was set at $p < 0.05$.

3 Results

Table 2 presents the percentage shares of short and normal-height children classified to each category of %FAT and BMI. No significant differences between short boys and girls in the occurrence of each category of %FAT (chi-square = 6.22; p = 0.10; Cramer's V = 0.18) or BMI (chi-square = 5.16;

Table 1 Somatic indices and selected birth parameters in short and normal-height children

Variable	Healthy short children				Healthy normal-height children			
	Girls		Boys		Girls		Boys	
	7–10 year (n = 41)	11–14 year (n = 44)	7–10 year (n = 41)	11–14 year (n = 56)	7–10 year (n = 179)	11–14 year (n = 127)	7–10 year (n = 186)	11–14 year (n = 117)
Sum of skinfolds (mm)	23.3 ± 8.6	30.8 ± 12.6	20.2 ± 10.5	23.9 ± 9.5	33.4 ± 16.4	40.8 ± 15.2	29.6 ± 15.5	37.2 ± 16.4
Body fat content (%)	15.4 ± 4.5	19.0 ± 5.1	12.5 ± 5.3	14.5 ± 5.0	18.2 ± 6.8	22.5 ± 7.6	16.9 ± 6.4	18.9 ± 5.8
Body mass index (kg/m^2)	15.1 ± 1.9	17.3 ± 2.9	15.5 ± 2.3	16.9 ± 2.1	16.9 ± 2.5	19.2 ± 3.3	17.2 ± 2.7	18.8 ± 2.8
Waist circumference (cm)	54.6 ± 4.7	59.8 ± 5.8	54.9 ± 4.8	61.5 ± 6.2	62.0 ± 7.4	69.8 ± 9.1	62.7 ± 7.8	71.2 ± 8.0
Waist-to-height ratio	0.44 ± 0.03	0.43 ± 0.04	0.44 ± 0.04	0.44 ± 0.04	0.45 ± 0.05	0.46 ± 0.06	0.46 ± 0.05	0.47 ± 0.05
Birth weight (g)[a]	2947 ± 398	3104 ± 408	3188 ± 594	3103 ± 526	–	–	–	–
Birth length (cm)[b]	51.5 ± 2.2	52.1 ± 3.6	52.7 ± 3.5	52.2 ± 4.3	–	–	–	–
Pregnancy length (weeks)[a]	37.9 ± 2.0	39.1 ± 1.8	38.8 ± 1.6	38.9 ± 2.7	–	–	–	–

Data are means ±SD
[a]numbers of children: girls 7–10 year, n = 16; girls 11–14 year, n = 24; boys 7–10 year, n = 10; boys 11–14 year, n = 28
[b]number of children: girls 7–10 year, n = 19; girls 11–14 year, n = 28; boys 7–10 year, n = 13; boys 11–14 year, n = 33

Table 2 Percentages of short and normal-height children stratified by the body fat percentage (%FAT) and the body mass index

	Healthy short children		Normal-height children	
	Girls (n = 85)	Boys (n = 97)	Girls (n = 306)	Boys (n = 303)
Body fat content				
Decreased	22.4 ± 4.5	32.0 ± 4.7	10.8 ± 1.8**	11.2 ± 1.8***
Normal	57.6 ± 5.4	59.7 ± 5.0	55.6 ± 2.8	60.4 ± 2.8
Increased (below obesity)	11.8 ± 3.5	5.2 ± 2.3	20.9 ± 2.3*	17.8 ± 2.2***
Obesity	8.2 ± 3.0	3.1 ± 1.8	12.7 ± 1.9	10.6 ± 1.8*
Body mass index				
Underweight	23.5 ± 4.6	23.7 ± 4.3	6.5 ± 1.4***	5.9 ± 1.4***
Normal	74.1 ± 4.8	67.0 ± 4.8	66.3 ± 2.7	66.0 ± 2.7
Overweight	2.4 ± 1.7	7.2 ± 2.6	18.0 ± 2.2***	19.8 ± 2.3**
Obesity	0.0 ± 0.0	2.1 ± 1.5	9.2 ± 1.7***	8.3 ± 1.6*

Data are means ±SE. Significantly different from the respective value in short children:*p < 0.05; **p < 0.01; ***p < 0.001. Percentages of children with decreased %FAT was significantly lower as compared to short children

Table 3 Pearson's correlation coefficients (*r*) between somatic indicators and birth parameters in short children aged 7–14 year (n = 93)

	Body fat content[a]	Body mass index[a]	Waist-to-height ratio
Birth weight	0.016	0.021	−0.034
Birth length	−0.015	−0.038	−0.081
Pregnancy length[b]	−0.036	−0.169	−0.209

[a]Evaluation based on z-scores; [b]n = 78

p = 0.16; Cramer's V = 0.15) were found. Increased %FAT or obesity were observed in nearly 14% of short children, whereas decreased body fat concerned 27% of short children. A similar share of short children with underweight was observed in case of BMI classification (24%). On the other hand, overweight and obesity were less frequent compared to %FAT estimations and occurred in about 6% of short children. In normal-height children decreased body fat or underweight were significantly less frequent compared to short children and amounted to about 11% and 6%, respectively. Increased %FAT was observed in 18–21% of normal-height children and BMI-based overweight and obesity in 18–20% and 8–9% of these children, respectively; the percentages were significantly higher than those observed in short children. Based on the cut-off points for WHtR that took into account age, sex and adiposity, abdominal obesity was found in about 4.5% of the short children and excessive abdominal adiposity was found in another 4.5% of these children. A nearly identical prevalence of abdominal obesity (5%; p = 0.82) was obtained when the popular criterion of WHtR ≥0.5 was used. In normal-height children, the number of children with WHtR ≥0.5 was threefold greater (p < 0.001) than that in short children and amounted to 18.7% in girls and to 17.4% in boys.

No significant associations between pregnancy length, birth weight, birth length, and the normalized values (z-scores) of %FAT, BMI, and WHtR were found among the short children. The correlation coefficients calculated for all short children were relatively low, ranging from −0.209 to 0.021 (Table 3).

Furthermore, no clear regularities were found between the weight-to-height proportions of short children and premature birth, low birth weight, or low birth length. Specifically, out of the nine short children who were born before the 37th week, five showed normal body mass, two were underweight, and two other were overweight at the time of the study. In turn, out of the eleven children with birth weight below the fifth centile, five had normal body mass, four were underweight, and two were either overweight or obese. Low birth length was observed in only two girls and one boy, two of these children were underweight, and one girl was overweight at the time of the study.

4 Discussion

The main findings of this study are a low overweight and obesity rate among short children and a high prevalence of decreased body fat or underweight reaching 25% of short children. We also could not substantiate the presence of any clear correlations between pregnancy length, birth weight and length with the actual body fat content, body mass index or waist-to-height ratio in children with height deficiency. Our findings can be considered of interest in view of many population studies indicating a high and steadily increasing prevalence of overweight and obesity among children. Namely, we found that the percentage of overweight or obese short children was 3–4 times lower than that observed in normal-height children in the Polish (Kułaga et al. 2011), European (Bodzsar and Zsakai 2014; Wijnhoven et al. 2013), or American (Ogden et al. 2016) populations. Conversely, the percentage of underweight short children was considerably greater than the 2–5% in other population studies (Wijnhoven et al. 2013; Kułaga et al. 2010) or the 10% reported in extreme cases depending on the children's age and societal wealth (de Onis et al. 2004; Wang et al. 2002), compared to normal-height children. Nevertheless, the results obtained in the present study are consistent with those reported in other studies conducted among short children. For instance, Tomaszewski et al. (2017) have observed a low prevalence of overweight and obesity in short children. Freedman et al. (2004b) have reported that short children, aged 5–8 year, show a lower total thickness of 10 skinfolds, about 10% lower body fat content, and lower BMI (15 *vs.* 19 kg/m^2) compared to peers with normal stature, which is similar to findings of the present study. However, sample size was relatively low in both aforementioned studies and the criterion of short stature was lenient (height below the 25th centile for a given age) in the study of Freedman et al. (2004b). A low prevalence of overweight and obesity, not exceeding 2% among children up to 10 year of age and 5–9% among youth below 17 year of age, has also been confirmed in another population study (Freedman et al. 2004a). A German study has found that short children have %FAT lower by 1–6% compared to children with normal stature, with about 8% of boys and 13% of girls of short stature having BMI below the 10th centile (Bosy-Westphal et al. 2009). In that study, overweight and obesity was diagnosed in about 6% of short children regardless of sex; the figure identical to that noted in the present study.

The observed low prevalence of overweight and obesity among short children could in part be related to the use of BMI in the evaluation (Freedman et al. 2005; Schaefer et al. 1998). Moreover, that fact that BMI correlates moderately with body height (Franklin 1999) may have led to an underestimation of BMI among short children (Wells 2001). Some authors suggest that when assessing the prevalence of overweight in short children, BMI should be compared to height age rather than chronological age (Bonthuis et al. 2013; National Kidney Foundation 2009). Apart from the characteristics of BMI itself, another factors that limit the accuracy of the diagnosis of overweight and obesity and consistency between the results in inter-population comparisons is the fact that researchers use different reference values and cut-off points. In the present study, reference values for the Polish population were used for the cut-off points of BMI (Tomaszewski et al. 2013), as the literature indicates that national, rather than international, standards are more accurate for the estimation of overweight and obesity (Chinn and Rona 2002; Reilly 2002). It has been previously shown that the cut-off points used in the present study (Tomaszewski et al. 2013, 2015) and the popular criteria proposed by Cole et al. (2000 and 2007) provide similar estimations of the prevalence of overweight and obesity. A substantial body composition difference between short and normal-height children found in the present study suggests that a low overweight and obesity rate in short children, along with low %FAT, is indeed a result of their small frame, rather than of any drawbacks stemming from the use of BMI or the cut-off points applied.

The assessments that involved the WHtR ≥0.5 criterion also indicate a lower prevalence of abdominal obesity in short children compared to normal-height children (5% *vs.* 18%) and to the general population. This suggests that abdominal obesity in short children is relatively low, and it is much lower than that in children and youth in the UK (McCarthy and Ashwell 2006), Spain (Schröder et al. 2014), and the US where nearly one in three has it (Li et al. 2006). However, a clear conclusion is difficult to achieve since WHtR has not been by far widely used to assess abdominal obesity in children with height deficit. The only exception is a most recent study by Tomaszewski et al. (2017) in which the prevalence of abdominal obesity has been estimated at 6.5% in healthy short children and 8–11% in children with growth hormone deficiency.

Low birth weight and premature birth have been demonstrated to cause growth disorders and, consequently, short stature and low body mass during adolescence. In children with low birth weight, the risk of short stature is nearly threefold greater at 7 year of age than in those with normal birth weight, and twofold greater than in adults (Greco et al. 1995). While prematurely born children show a higher waist circumference as newborns, at 5 year of age they are slimmer compared to the reference group (Roswall et al. 2012). On the other hand, malnutrition during pregnancy and early childhood has been shown to correlate with increased risk of obesity and comorbidities during adolescence and adulthood (Monasta et al. 2010; Rugholm et al. 2005). The present study found no significant correlations between pregnancy length, birth weight, and birth length and %FAT, BMI, or WHtR among the children with short stature. Nearly half of the children who were born prematurely or had low birth weight showed a normal body mass. Other studies conducted among children with normal body height also provide unclear results (Monasta et al. 2010; Apfelbacher et al. 2008; Hirschler et al. 2008; Rugholm et al. 2005). The literature consistently indicates that the risk of metabolic disorders and obesity among children with low birth weight is related to a higher rate of weight gain shortly after birth (Fabricius-Bjerre et al. 2011). For this reason, the occurrence of obesity and metabolic disorders at a later age is assumed to primarily result from the aforementioned adaptive mechanisms, rather than low birth weight (Yu et al. 2011). Plausibly, these adaptive factors, not taken into account in the present study, along with the specificity of a given population with short stature, are what determines the lack of clear associations of birth weight and adiposity with body mass indicators.

Apart from birth parameters, dietary factors also play a role in shaping the weight-to-height proportion in children with short stature. These children are considerably underweight and suffer from being 'poor and picky eaters' (Wudy et al. 2005). This may, to some extent, explain a lower prevalence of overweight and obesity among short children and youth, and a relatively high prevalence of insufficient %FAT and underweight observed in the present study. It cannot be excluded that a slower growth rate in persons with short stature make them intake a small amount of food when their energy stocks are sufficient. A correct nutrition of short children should be a subject of concern for physicians, parents, and, at an older age, children themselves. On the other side, excessive focus on a child's body mass may negatively affect his self-esteem and perception of the body (Kohn et al. 2006).

Apart from its strengths that stems from a relatively big sample size, a narrow children's age range, and the inclusion of birth parameters of short children, the present study has certain shortcomings. They are a lack of data on the children's development rate, psychosocial status, diet, and physical activity; the elements that could have a bearing on the weight-to-height proportion and on the cause of insufficient body mass observed in short children. Furthermore, incompleteness of data on pregnancy length and children's birth weight and length prevents drawing conclusions concerning the association between birth parameters and somatic traits in short children.

In conclusion, the study findings demonstrate that the prevalence of obesity, including abdominal obesity, is low (<6%) among short children and youth, especially compared to the normal-height children and population estimates. The occurrence of underweight and insufficient amount of fat tissue, which were observed in one out of four children, should be considered particularly disturbing in the context of a correct development of short children. The obtained results do not indicate any clear association between pregnancy length and birth weight and length, and the actual %FAT, BMI, or WHtR in short children. Further research on larger population samples is required to assess the viability of BMI and WHtR in the evaluation of children and adolescents with short stature and to verify the commonly used cut-off points, designed for the general population.

Acknowledgments This work was partly supported by grant no. AWF-DS.153 of the Polish Ministry of Science and Higher Education.

Conflicts of Interest The authors declare no conflicts of interest in relation to this article.

References

Apfelbacher CJ, Loerbroks A, Cairns J, Behrendt H, Ring J, Kramer U (2008) Predictors of overweight and obesity in five to seven year-old children in Germany: results from cross-sectional studies. BMC Public Health 8:171

Bierich JR (1992) Constitutional delay of growth and adolescence. Bailliere Clin Endocrinol Metab 6:573–588

Bodzsar EB, Zsakai A (2014) Recent trends in childhood obesity and overweight in the transition countries of Eastern and Central Europe. Ann Hum Biol 41:263–270

Bonthuis M, Jager KJ, Abu-Hanna A, Verrina E, Schaefer F, van Stralen KJ (2013) Application of body mass index according to height-age in short and tall children. PLoS One 8:e72068

Bosy-Westphal A, Plachta-Danielzik S, Dörhöfer RP, Müller MJ (2009) Short stature and obesity: positive association in adults but inverse association in children and adolescents. Br J Nutr 102:453–461

Browning LM, Hsieh SD, Ashwell M (2010) A systematic review of waist-to-height ratio as a screening tool for the prediction of cardiovascular disease and diabetes: 0.5 could be a suitable global boundary value. Nutr Res Rev 23:247–269

Chinn S, Rona RJ (2002) International definitions of overweight and obesity for children: a lasting solution? Ann Hum Biol 29:306–313

Cole TJ, Bellizzi MC, Flegal KM, Dietz WH (2000) Establishing a standard definition for child overweight and obesity worldwide: international survey. BMJ 320:1240–1243

Cole TJ, Flegal KM, Nicholls D, Jackson AA (2007) Body mass index cut offs to define thinness in children and adolescents: international survey. BMJ 335:166–167

de Onis M, Blössner M, Borghi E, Frongillo EA, Morris R (2004) Estimates of global prevalence of childhood underweight in 1990 and 2015. JAMA 291:2600–2606

Dobosz J (2012) Physical development of school-age children and adolescents. Percentile charts. University of Physical Education, Warsaw. https://demo.ncbkf.pl/pliki/dl/publikacje/Kondycja%20fizyczna_siatki%20centylowe.pdf. Accessed 9 July 2017

Fabricius-Bjerre S, Jensen RB, Færch K, Larsen T, Mølgaard C, Michaelsen KF, Vaag A, Greisen G (2011) Impact of birth weight and early infant weight gain on insulin resistance and associated cardiovascular risk factors in adolescence. PLoS One 6:e20595

Franklin MF (1999) Comparison of weight and height relations in boys from 4 countries. Am J Clin Nutr 70:157–162

Freedman DS, Khan LK, Serdula MK, Dietz WH, Srinivasan SR, Berenson GS (2004a) Interrelationships among childhood BMI, childhood height, and adult obesity: the Bogalusa Heart Study. Int J Obes Relat Metab Disord 28:10–16

Freedman DS, Thornton JC, Mei Z, Wang J, Dietz WH, Pierson RN Jr, Horlick M (2004b) Height and adiposity among children. Obes Res 12:846–853

Freedman DS, Wang J, Maynard LM, Thornton JC, Mei Z, Pierson RN, Dietz WH, Horlick M (2005) Relation of BMI to fat and fat-free mass among children and adolescents. Int J Obes 29:1–8

Greco L, Power C, Peckham C (1995) Adult outcome of normal children who are short or underweight at age 7 years. BMJ 310:696–700

Hall JG, Allanson JE, Gripp KW, Slavotinek AM (2007) Handbook of physical measurements, 2nd edn. Oxford University Press, Oxford

Hirschler V, Bugna J, Roque M, Gilligan T, Gonzalez C (2008) Does low birth weight predict obesity/overweight and metabolic syndrome in elementary school children? Arch Med Res 39:796–802

Kohn M, Rees JM, Brill S, Fonseca H, Jacobson M, Katzman DK, Loghmani ES, Neumark-Sztainer D, Schneider M (2006) Preventing and treating

adolescent obesity: a position paper of the Society for Adolescent Medicine. J Adolesc Health 38:784–787
Kułaga Z, Litwin M, Tkaczyk M, Różdżyńska A, Barwicka K, Grajda A, Świąder A, Gurzkowska B, Napieralska E, Pan H (2010) The height-,weight-, and BMI-for-age of Polish school-children and adolescents relative to international and local growth references. BMC Public Health 10:109
Kułaga Z, Litwin M, Tkaczyk M, Palczewska I, Zajączkowska M, Zwolińska D, Krynicki T, Wasilewska A, Moczulska A, Morawiec-Knysak A, Barwicka K, Grajda A, Gurzkowska B, Napieralska E, Pan H (2011) Polish 2010 growth references for school-aged children and adolescents. Eur J Pediatr 170:599–609
Li C, Ford ES, Mokdad AH, Cook S (2006) Recent trends in waist circumference and waist-height ratio among US children and adolescents. Pediatrics 118:e1390–e1398
McCarthy HD, Ashwell M (2006) A study of central fatness using waist-to-height ratios in UK children and adolescents over two decades supports the simple message -'keep your waist circumference to less than half your height'. Int J Obes 30:988–992
Monasta L, Batty GD, Cattaneo A, Lutje V, Ronfani L, Van Lenthe FJ, Brug J (2010) Early-life determinants of overweight and obesity: a review of systematic reviews. Obes Rev 11:695–708
National Kidney Foundation (2009) KDOQI clinical practice guideline for nutrition in children with CKD: 2008 update. Am J Kidney Dis 53:S1–S124
Ogden CL, Carroll MD, Lawman HG, Fryar CD, Kruszon-Moran D, Kit BK, Flegal KM. (2016) Trends in obesity prevalence among children and adolescents in the United States, 1988–1994 through 2013–2014. JAMA 315(21):2292–2299
Pasquino AM, Albanese A, Bozzola M, Butler GE, Buzi F, Cherubini V, Chiarelli F, Cavallo L, Drop SL, Stanhope R, Kelnar CJ (2001) Idiopathic short stature. J Pediatr Endocrinol Metab 14:967–974
Reilly JJ (2002) Assessment of childhood obesity: national reference data or international approach? Obes Res 10:838–840
Roswall J, Karlsson AK, Allvin K, Tangen GA, Bergman S, Niklasson A, Alm B, Dahlgren J (2012) Preschool children born moderately preterm have increased waist circumference at two years of age despite low body mass index. Acta Paediatr 101:1175–1181
Rugholm S, Baker JL, Olsen LW, Schack-Nielsen L, Bua J, Sørensen TI (2005) Stability of the association between birth weight and childhood overweight during the development of the obesity epidemic. Obes Res 13:2187–2194
Schaefer F, Georgi M, Wühl E, Schärer K (1998) Body mass index and percentage fat mass in healthy German schoolchildren and adolescents. Int J Obes Relat Metab Disord 22:461–469
Schröder H, Ribas L, Koebnick C, Funtikova A, Gomez SF, Fíto M, Perez-Rodrigo C, Serra-Majem L (2014) Prevalence of abdominal obesity in Spanish children and adolescents. Do we need waist circumference measurements in pediatric practice? PLoS One 9: e87549
Slaughter MH, Lohman TG, Boileau RA, Horswill CA, Stillman RJ, Van Loan MD, Bemben DA (1988) Skinfold equations for estimation of body fatness in children and youth. Hum Biol 60:709–723
Stupnicki R, Tomaszewski P, Milde K, Głogowska J, Popławska H (2013) Waist-to-height ratio (WTHR): proposed approach to establishing reference values for children and youths. Papers Anthropol 22:214–224
Thibault H, Souberbielle JC, Taieb C, Brauner R (1993) Idiopathic prepubertal short stature is associated with low body mass index. Horm Res 40:136–140
Tomaszewski P, Stupnicki R, Milde K (2013) Body mass index – proposed norms for children and youths. Papers Anthropol 22:203–213
Tomaszewski P, Żmijewski P, Milde K, Sienkiewicz-Dianzenza E (2015) Weight-height relationships and central obesity in 7-year-old to 10-year-old Polish urban children: a comparison of different BMI and WHtR standards. J Physiol Anthropol 34, 34
Tomaszewski P, Milde K, Majcher A, Pyrżak B, Tiryaki-Sonmez G, Schoenfeld BJ (2017) Body mass disorders in healthy short children and in children with growth hormone deficiency. Adv Exp Med Biol. https://doi.org/10.1007/5584_2017_65
Voss LD, Mulligan J, Beets PR, Wilkin TJ (1992) Poor growth in school entrants as an index of organic disease: the Wessex growth study. BMJ 305:1400–1402
Wang Y, Monteiro C, Popkin BM (2002) Trends of obesity and underweight in older children and adolescents in the United States, Brazil, China, and Russia. Am J Clin Nutr 75:971–977
Wells JC (2001) A critique of the expression of paediatric body composition data. Arch Dis Child 85:67–72
WHO (2006) Multicentre growth reference study group. WHO child growth standards: length/height-for-age, weight-for-age, weight-for-length, weight-for-height and body mass index-for-age: methods and development. World Health Organization, Geneva
Wijnhoven TM, van Raaij JM, Spinelli A, Rito AI, Hovengen R, Kunesova M, Starc G, Rutter H, Sjöberg A, Petrauskiene A, O'Dwyer U, Petrova S, Farrugia Sant'Angelo V, Wauters M, Yngve A, Rubana IM, Breda J (2013) WHO European childhood obesity surveillance initiative 2008: weight, height and body mass index in 6-9-year-old children. Pediatr Obes 8:79–97
Wudy SA, Hagemann S, Dempfle A, Ringler G, Blum WF, Berthold LD, Alzen G, Gortner L, Hebebrand J (2005) Children with idiopathic short stature are poor eaters and have decreased body mass index. Pediatrics 116:e52–e57
Yu ZB, Han SP, Zhu GZ, Zhu C, Wang XJ, Cao XG, Guo XR (2011) Birth weight and subsequent risk of obesity: a systematic review and meta-analysis. Obes Rev 12:525–542

Advs Exp. Medicine, Biology - Neuroscience and Respiration (2018) 38: 71–80
DOI 10.1007/5584_2017_107

Published online: 5 October 2017

Neutrophil Gelatinase-Associated Lipocalin: A Biomarker for Early Diagnosis of Urinary Tract Infections in Infants

Grażyna Krzemień, Małgorzata Pańczyk-Tomaszewska, Dominika Adamczuk, Iwona Kotuła, Urszula Demkow, and Agnieszka Szmigielska

Abstract

Early diagnosis of urinary tract infection (UTI) is challenging in infants due to unspecific symptoms, difficulty in urine collection and possible contamination. The aim of this study was to assesses the usefulness of serum and urine neutrophil gelatinase-associated lipocalin (sNGAL and uNGAL, respectively) in the diagnosis of febrile and non-febrile UTI in infants. This prospective observational study enrolled 66 infants with the first episode of UTI and 18 healthy controls. At the time of enrollment, sNGAL, uNGAL, urinalysis, urine culture, white blood cell count (WBC), C-reactive protein (CRP), erythrocyte sedimentation rate (ESR), procalcitonin (PCT), and serum creatinine (sCr) were assessed. We found that, on average, both sNGAL and uNGAL levels were significantly higher in febrile UTI, compared to non-febrile UTI and controls. In turn, the mean sNGAL level, but not uNGAL, was significantly higher in the non-febrile UTI group compared to controls. sNGAL positively correlated with WBC, CRP, ESR and PCT, and uNGAL with CRP and leukocyturia. The receiver operating curves (ROC) demonstrate that the optimum cut-off of 76.2 ng/ml for sNGAL (sensitivity 92.9%, specificity 94.4%, and the area under the curve (AUC) of 0.98) and of 42.2 ng/ml for uNGAL (sensitivity 73.8%, specificity 72.2%, and AUC of 0.76) for diagnosing febrile UTI and 39.0 ng/ml for sNGAL (sensitivity 83.3%, specificity 55.6%, and AUC of 0.70) for diagnosing non-febrile UTI. In conclusion, serum NGAL is an excellent marker for the early diagnosis of febrile UTI, with sensitivity and specificity higher than those of urine NGAL. Diagnostic sensitivity of serum NGAL is smaller in non-febrile infants suffering from UTI, and urine NGAL is not useful for this purpose at all.

G. Krzemień, M. Pańczyk-Tomaszewska, D. Adamczuk, and A. Szmigielska (✉)
Department of Pediatrics and Nephrology, Warsaw Medical University, 63A Żwirki and Wigury Street, 02-091 Warsaw, Poland
e-mail: agnieszka.szmigielska@wum.edu.pl

I. Kotuła and U. Demkow
Department of Laboratory Diagnostics and Clinical Immunology of Developmental Age, Warsaw Medical University, Warsaw, Poland

Keywords
Biomarkers · Children · Lipocalin · Neutrophils · Receiver operating curve · Urinanalysis · Urine culture · Urinary tract infection

1 Introduction

Urinary tract infection (UTI) is one of the most common bacterial infection in children (Stein et al. 2015). Urine culture is the gold standard for diagnosing UTI, but culture results require at least 2 d for a complete identification of bacteria (Ylmaz et al. 2009). Therefore, results of urinalysis, such as the presence of leukocyturia, and nitrite and leukocyte esterase, are used to diagnose UTI in clinical practice, but low sensitivity and specificity of this analysis is an issue (Kim et al. 2014). Difficulty in urine collection or improperly collected specimen may lead to false positive results of urine culture and urinalysis in infants (Hatipoglu et al. 2011). Additionally, leukocyturia in infants may be a sign of a wide range of inflammatory changes, including perineum, vaginitis, diarrhoea, synechia, or phimosis. Sterile leukocyturia may also occur in noninfectious conditions such as urolithiasis, congenital abnormalities of kidney and urinary tract (CAKUT), recent history of urological intervention, dehydratation, and high fever of external causes (Szmigielska and Krzemień 2017; AAP 2011).

Early diagnosis of UTI in infants is especially important, because a delay in diagnosis and treatment can result in severe renal parenchymal involvement and can increase the risk of chronic kidney damage, including renal failure (Ghasemi et al. 2016; Sim et al. 2015). Therefore, novel blood and urine biomarkers are needed for an early diagnosis of UTI in children (Kim et al. 2014; Yim et al. 2014).

Neutrophil gelatinase-associated lipocalin (NGAL) is an important component of the innate antimicrobial immune system (Ylmaz et al. 2009). The main bacteriostatic effect of NGAL on gram-negative bacteria is binding and sequestration of iron-loaded bacterial siderophores, which prevents bacterial iron uptake and growth. NGAL also modulates neutrophil-related processes such as maturation, adhesion, phagocytosis, or bacterial killing and it acts as chemoattractant for neutrophils. In healthy individuals, NGAL is expressed at very low concentrations in neutrophils and in epithelial cells of several tissues including kidney, liver, lungs, and colon (Nasioudis and Witkin 2015). In acute kidney injury, NGAL is massively released into the blood and urine from nephron epithelium (Singer et al. 2013). NGAL also is an acute phase reactant and may be released from activated neutrophils and other immune cells (Kim et al. 2014). Usefulness of plasma and urine NGAL (pNGAL and uNGAL) measurements for diagnosing chronic kidney damage and different bacterial infection has been reported in various studies (Filho et al. 2017; Singer et al. 2013). Recent data show that serum NGAL (sNGAL), alongside pNGAL and uNGAL assay, can be used for prediction of UTI (Yim et al. 2014; Ylmaz et al. 2009), acute pyelonephritis (Kim et al. 2017; Nickavar et al. 2016), or renal scarring (Rafiei et al. 2015). Clinical studies assessing the usefulness of NGAL in UTI diagnosis in children have been performed in a small number of patients, are single-center studies, and the results are contentious (Ghasemi et al. 2016; Lee et al. 2015; Kim et al. 2014). Therefore, the aim of the present study was to assess the usefulness of sNGAL and uNGAL in the early diagnosis of febrile and non-febrile UTI in infants.

2 Methods

This prospective observational study was performed in the Department of Pediatric and Nephrology of Warsaw Medical University in Warsaw, Poland of over a period of 3 years. The study was approved by a local Bioethics Committee for Human Research. All parents gave their informed consent. Eighty four children (42 males, 42 females, mean age 5.8 ±3.4 months, range 1–12 months) were enrolled in the study: 66 infants with a first episode of UTI and 18 healthy infants as a control group. Children with other infections, symptoms of chronic kidney damage, chronic kidney damage, liver dysfunction, evidence of obstructive nephropathy, and moderate- or severe-grade of vesicoureteral reflux were excluded from the study. Based on the clinical manifestation and laboratory tests, children were divided into three groups: febrile UTI, non-febrile UTI, and healthy controls. Urinalysis, urine culture, white blood cell count (WBC), C-reactive protein (CRP), erythrocyte sedimentation rate (ESR), procalcitonin (PCT), and serum creatinine (sCr) were performed on admission. Urine samples were collected *via* a catheter or from a midstream urine. Leukocyturia was defined as >10 WB *per* high power field in the urine sediment. Significant bacteriuria was defined as growth of a single pathogen $\geq 10^5$/colony-forming units *per* ml urine. Normal values of the indices evaluated in the local laboratory were as follows: WBC $\leq$ 18.000/mm^3, CRP $\leq$ 1.0 mg/dl, ESR $\leq$ 15 mm/h, PCT $\leq$ 0.05 ng/ml, and sCr $\leq$ 0.4 mg/dl. The UTI diagnosis was based on significant bacteriuria and leukocyturia. Febrile UTI was defined as the presence of fever (≥38.0 °C). Serum and urine samples collected for the measurement of NGAL were obtained at the time of admission. They were immediately centrifuged and stored at –80 °C until further analysis.

Urine and serum NGAL content was determined using enzyme-linked immunoadsorbent assay (ELISA) kit (Human Lipocalin-2/NGAL; Cat no: RD 191102200R; BioVendor Laboratory Medicine Inc., Brno, Czech Republic). The detection range for NGAL was 0.3–10 ng/ml. Urine NGAL to urine creatinine ratio (uNGAL/uCr) was calculated to normalize samples and was expressed as ng/mg. Renal ultrasound was performed after admission. Voiding cystourethrography (VCUG) was performed 2–3 weeks after treatment of UTI in children with any abnormal finding in renal ultrasound.

2.1 Statistical Evaluation

The Liliefors and Shapiro-Wilk tests were used to determine whether the data were normally distributed. Continuous data were presented as means ±SD or median and interquartile ranges (IQ). Categorical data were presented as numbers and percentiles. One-way ANOVA or a non-parametric Kruskal-Wallis test were used to compare continuous variables between the three groups. A *post hoc* test was used to compare subgroups. A chi-squared test was used to compare categorical variables. Correlations between variables were evaluated by linear regression analysis or Sperman's rank correlation. The receiver operating curves (ROC) were made and areas under the curves (AUC) were calculated, including 95% confidence intervals (CI), to determine sensitivity and specificity, and the optimal cut-off values of sNGAL, uNGAL, and uNGAL/uCr ratio to predict UTI. A p-value <0.05 defined statistically significant differences. Statistical evaluation was performed using a commercial Statistica package v11.0 for Windows (StatSoft; Tulssa, OK).

3 Results

Febrile UTI was diagnosed in 42 (64%) and non-febrile UTI in 24 (36%) children. The control group consisted of 18 children without UTI. *Escherichia coli* was isolated in 60 (91%) children with UTI, and other bacteria in 6 (9%) children (*Klebsiella sp.* in two, *Enterococcus sp.* in two, and *Proteus sp.* in two children). Indications

Table 1 Demographic, laboratory, and imaging findings in children with febrile and non-febrile urinary tract infection (UTI), and in controls

Variable	Group			p		
	Febrile (1) (n = 42)	Non-febrile (2) (n = 24)	Controls (3) (n = 18)	1 *vs.* 2	1 *vs.* 3	2 *vs.* 3
Age (month)	7.1 ± 3.4	3.9 ± 2.6	5.2 ± 3.0	<0.001	<0.05	ns
Male, *n* (%)	25 (59.5)	9 (37.5)	8 (44.4)	ns	ns	ns
sNGAL (ng/ml)	159.6 ± 63.9	60.6 ± 19.9	46.3 ± 22.2	<0.0001	<0.0001	<0.05
uNGAL (ng/ml)	67.5 ± 34.8	48.0 ± 41.1	38.0 ± 32.3	<0.05	<0.01	ns
uNGAL/uCr	172. 0	195.2	178.3	ns	ns	ns
IQ	116.5 – 267.9	85.8 – 371.5	44.1 – 338.0			
↑ WBC, *n* (%)	25 (59)	5 (19)	0	<0.001	<0.0001	<0.05
↑ CRP, *n* (%)	42 (100)	14 (58)	0	<0.001	<0.0001	<0.001
↑ ESR, *n* (%)	36 (100)	14 (64)	0	<0.001	<0.0001	<0.0001
↑ PCT, *n* (%)	42 (100)	14 (58)	0	<0.001	<0.0001	<0.001
WBC (mm^3)	20.4 ± 6.2	14.5 ± 4.3	11.2 ± 2.5	<0.001	<0.0001	<0.01
CRP (mg/dl)	8.8	1.5	0.5	<0.0001	<0.0001	<0.0001
IQ	5.7 – 17.3	0.6 – 2.9	0.5 – 0.5			
ESR (mm/h)	72.0 ± 31.4	27.3 ± 18.1	9.1 ± 4.3	<0.0001	<0.0001	<0.05
PCT (ng/ml)	0.96	0.08	0.05	<0.0001	<0.0001	<0.01
IQ	0.34–2.33	0.05–0.09	0.05–0.05			
Leukocyturia (uWBC/HPF)	100.0	55.0	1.0	ns	<0.0001	<0.0001
IQ	30.0 – 100.0	27.5 – 100.0	0.0 – 3.0			
CAKUT, *n* (%)	5 (11)	5 (21)	4 (22)	ns	ns	ns

Data are presented as means ± SD, median with interguartile range (IQ), or number (%); *sNGAL* serum neutrophil gelatinase-associated lipocalin, *uNGAL* urine neutrophil gelatinase-associated lipocalin, *uNGAL/uCr* urine NGAL to urine creatinine ratio, *WBC* white blood cell count, *CRP* serumC-reactive protein, *ESR* erythrocyte sedimentation rate (performed in 76 children), *PCT* procalcitonin, ↑ number of patients with elevated values, *uWBC/HPF* urine WBC/high power field, *CAKUT* congenital anomaly of kidney and urinary tract

for urinalysis and urine culture in the control group included the following: failure to thrive (three), prolonged jaundice (two), and abnormalities in the urinary tract unraveled in ultrasonography (four). Routine urine tests were performed in 9 children. In all children in the control group, urine culture was negative and urinalysis was normal. sCr was normal in all children evaluated in the study.

A comparison of demographic, laboratory and imaging findings in children with febrile and non-febrile UTI, and in healthy controls are presented in Table 1. The children with febrile UTI were significantly older than those with both non-febrile UTI and healthy controls. There was no significant difference in age in children with non-febrile UTI compared to controls. Gender distribution did not differ between the groups. The mean sNGAL and uNGAL levels were significantly higher in the febrile UTI group than those in both non-febrile UTI and controls. The mean sNGAL, but not uNGAL, was significantly higher in the non-febrile UTI than that in controls. There were no significant differences in the median uNGAL/uCr ratio among the groups of children.

Elevated values of WBC, CRP, ESR i PCT were present significantly more often in the febrile UTI than those in the non-febrile UTI and control groups, and in the non-febrile UTI than in the control group. WBC, CRP, ESR and PCT were significantly higher in the febrile UTI than those in the non-febrile UTI and control groups, and in the non-febrile UTI than in the control group. No significant differences in the urine WBC count were documented between children with febrile and non-febrile UTI. Based on renal ultrasound and voiding

Table 2 Correlations between sNGAL, uNGAL, uNGAL/uCr ratio and age, systemic inflammatory markers, leukocyturia

	sNGAL		uNGAL		uNGAL/uCr	
Variable	*r*	p	*r*	p	*r*	p
Age	0.33	<0.01	–	–	−0.26	<0.05
uNGAL	–	–	–	–	0.53	<0.001
WBC	0.70	<0.001	–	–	–	–
CRP	0.67	<0.001	0.30	<0.01	–	–
ESR	0.59	<0.001	–	–	–	–
PCT	0.53	<0.001	–	–	–	–
Leukocyturia	–	–	0.36	<0.01	–	–

sNGAL serum neutrophil gelatinase-associated lipocalin, *uNGAL* urine neutrophil gelatinase-associated lipocalin, *uNGAL/uCr* urine NGAL to urine creatinine ratio, *WBC* white blood cell count, *CRP* C-reactive protein, *ESR* erythrocyte sedimentation rate, *PCT* procalcitonin

Table 3 Area under the curve (AUC), optimal cut-off value, sensitivity and specificity for sNGAL, uNGAL, and uNGAL/uCr ratio for the diagnosis of febrile and non-febrile UTI

Variable	AUC	Z-test	p	Cut-off	Sensitivity	Specificity
Febrile UTI						
sNGAL	0.98 (95% CI 0.96–1.00)	39.5	<0.00001	76.2	92.9%	94.4%
uNGAL	0.76 (95% CI 0.61–0.89)	3.43	<0.001	42.2	73.8%	72.2%
uNGAL/uCr	0.55 (95% CI 0.38–0.73)	0.58	ns	–	–	–
Non-febrile UTI						
sNGAL	0.70 (95% CI 0.54–0.87)	2.38	<0.05	39.0	83.3%	55.6%
uNGAL	0.56 (95% CI 0.39–0.74)	0.71	ns	–	–	–
uNGAL/uCr	0.55 (95% CI 0.37–0.73)	0.55	ns	–	–	–

sNGAL serum neutrophil gelatinase-associated lipocalin, *uNGAL* urine neutrophil gelatinase-associated lipocalin, *uNGAL/uCr* urine NGAL to urine creatinine ratio, *UTI* urinary tract infection

cystouretrography, CAKUT was diagnosed in 14 (16.7%) children, including five with febrile UTI (four children with vesicoureteral reflux Grade 2 and one child with duplex kidney), five with non-febrile UTI (four children with mild hydronephrosis and one child with duplex kidney), and in four children mild hydronephrosis from the control group. There were no significant differences in frequency of CAKUT among the groups of children.

Correlations between sNGAL, uNGAL, uNGAL/uCr, one side and age, systemic inflammatory markers, and leukocyturia on the other side, are presented in Table 2. Positive correlations were present between uNGAL and uNGAL/uCr. sNGAL positively correlated with age, WBC, CRP, ESR, and PCT. uNGAL positively correlated only with CRP and leukocyturia. sNGAL correlation with age was lower than that with inflammatory markers. uNGAL/uCr was inversely correlated with age, and uCr was positively correlated with age. Gender had no appreciable influence on sNGAL or uNGAL levels (data not shown).

Area under the curve calculated with the ROC method, optimal cut-off value, sensitivity and specificity for sNGAL, uNGAL, and uNGAL/uCr for diagnosing of febrile and non-febrile UTI are shown in Table 3. ROC analysis demonstrated that sNGAL had the greatest AUC value of 0.98 for diagnosing febrile UTI, followed by uNGAL of 0.76 for diagnosing febrile UTI, and sNGAL of 0.70 for diagnosing non-febrile UTI.

4 Discussion

In infants, early diagnosis of UTI is often challenging, because of nonspecific symptoms of infection, difficulty in urine collection and a high risk of contamination (Lee et al. 2013; Tullus 2011). Inadequately collected specimens can lead to misdiagnosis and to unnecessary antibiotic treatment in children with false positive changes (Krzemień et al. 2014; Hatipoglu et al. 2011). Thus, there is a need for sensitive and specific laboratory tests for accurate diagnose of UTI in infants and young children. Ichino et al. (2009) have demonstrated in an experimental study that uNGAL increases in the early stage of UTI, suggesting that uNGAL might be used for predicting UTI. Since that time, usefulness of blood and urine NGAL assessment in UTI have been investigated in animal and human studies (Forster et al. 2017; Kim et al. 2017; Urbschat et al. 2014).

The main source of blood NGAL in UTI are activated neutrophiles in the infected kidney and activated peripheral neutrophiles in case of neutrophilia (Lee et al. 2015; Kim et al. 2014). In patients with severe form of acute pyelonephritis and concomitant chronic kidney damage, NGAL is released from injured nephron epithelia into blood and urine (Singer et al. 2013). Urine NGAL derives mainly from activated neutrophiles present in urine and in the infected kidney, less from injured nephron epithelia (Decavele et al. 2011).

In the present study we assessed sNGAL and uNGAL in a cohort of infants with UTI and found significantly higher sNGAL and uNGAL levels in infants with febrile UTI compared to non-febrile UTI and healthy controls, and a significantly higher sNGAL level in infants with non-febrile UTI compared to the controls. Ylmaz et al. (2009) have demonstrated that uNGAL increases in children diagnosed with UTI. Those authors have evaluated 60 children with symptomatic UTI and reported that uNGAL and uNGAL/uCr are significantly higher in the UTI group than in controls. They have also determined the optimum cut-off value of 20 ng/ml for uNGAL (sensitivity 97%, specificity 76%, and AUC 0.98) and of 30 for uNGAL/uCr (sensitivity 98%, specificity 76%, and AUC 0.99) for diagnosing UTI and concluded that uNGAL and uNGAL/uCr are sensitive markers for early UTI diagnose in the absence of chronic kidney damage or chronic kidney damage. In a similar study, Yim et al. (2014) have investigated 73 children with febrile UTI and also found that uNGAL and uNGAL/uCr are significantly higher in the UTI group than in controls. Using the optimum cut-off value of 23.95 ng/ml for uNGAL (sensitivity 82.4%, specificity 83.6%, and AUC 0.89) and of 276.5 for uNGAL/uCr (sensitivity 76%, specificity 90%, and AUC 0.90), they noted smaller sensitivity, but greater specificity of both markers for an early UTI diagnosis than those in the of Ylmaz et al. (2009). It is worth noting that the optimum cut-off values of uNGAL normalized to uCr in both studies are different. These differences are due most likely to different age of children in both studies. Children studied by Yim et al. (2014) were younger than those studied by Ylmaz et al. (2009). Additionally, it is known that concentration of uNGAL is high and concentration of uCr is low during infancy.

Another study has demonstrated a significantly higher level of uNGAL/Cr in 33 children with febrile UTI than in controls (Lee et al. 2015). With the best cut-off value of 108 ng/mg, uNGAL/uCr shows a good diagnostic profile for UTI (sensitivity 97%, specificity 80.6%, and AUC 0.91). A study by Urbschat et al. (2014), performed in 59 adults with clinical symptoms of upper and lower UTI, has revealed a significantly higher uNGAL level in patients with UTI than in controls. Arambašić et al. (2016) have noted that uNGAL can be useful for the differentiation of cystitis from febrile states other than UTI. Hatipoglu et al. (2011) have assessed a complex of matrix metalloproteinase-9 (MMP-9) and NGAL in 71 children with lower UTI and in 37 with an asymptomatic contamination. Those authors report that the uMMP-9/NGAL/uCr level is significantly higher in the UTI group, but it is unelevated in asymptomatic children with contaminated urine and in healthy controls. Using the optimum cut-off value 0.08 ng/mg for uMMP-9/NGAL/Cr (sensitivity 98.6%, specificity 97.3%, and AUC 0.99) they found higher sensitivity and specificity values of this marker for the diagnosis of UTI, than those reported by Lee et al. (2015), Yim et al. (2014), and Ylmaz et al. (2009) for uNGAL and uNGAL/uCr.

According to the reports above outlined, uNGAL, uNGAL/Cr, and uMMP-9/uNGAL/uCr demonstrate high sensitivity, specificity, and AUC value for the diagnosis of UTI. The increases in those markers were associated with UTI severity and they dropped significantly following antibiotic treatment. The markers were useful for diagnosing and managing febrile UTI in children, and also for the differentiation between contamination and infection. In Ylmaz et al. (2009) and Hatipoglu et al.'s (2011) opinions, when using the optimum cut-off value for the markers, treatment can be started even before the results of urine culture become available. In patients with persistently elevated uNGAL/uCr, there is a risk of inappropriate treatment of febrile UTI. Therefore, Yim et al. (2014) have recommended dimercaptosuccinic acid (DMSA) scans for these patients to exclude renal scars. There is a consensus that uNGAL and uNGAL/uCr levels cannot replace the urinalysis and urine culture in UTI diagnosis, but measurements of these markers can improve the management algorithms.

In the present study we found lower sensitivity and specificity of uNGAL (73.8% and 72.2%, respectively) for diagnosing febrile UTI than those in the studies above outlined. We also found that uNGAL was not useful for diagnosing non-febrile UTI. In contrast, results of some other studies have demonstrated high sensitivity of uNGAL normalized for uCr for predicting UTI. Normalization of urine biomarkers for uCr is a common practice, although it is a debatable approach (Bennett et al. 2015; McWilliam et al. 2014). In chronic kidney damage, the absolute values of biomarkers can be misleading due to a difference in urinary flow rate. Likewise, normalized values of biomarkers can be misleading due to variations in uCr excretion (Waikar et al. 2010). Forster et al. (2017) have suggested that in patients with UTI, uNGAL normalization for uCr is an imperfect standardization due to distinctly different biologies of uNGAL and uCr. In patients with UTI, uNGAL is released from injured tubular cells or neutrophiles present in urine, whereas uCr is depended on the glomerular filtration rate. In present study, uNGAL/uCr was inversely associated with age. Normalization of uNGAL for uCr did not improve the prognostic value of this biomarker in UTI diagnosis. This finding is in line with those from other studies (Rafiei et al. 2015; Kim et al. 2014; Yim et al. 2014).

In children, the measurement of urine biomarkers is preferred over blood tests as it is an easier and less stressful procedure. Therefore, only a few investigators assessed sNGAL or pNGAL in children with UTI. Most of them evaluated the usefulness of NGAL measurements to predict acute pyelonephritis in patients with febrile UTI (Kim et al. 2017; Sim et al. 2015). Seo et al. (2014) have evaluated 47 infants with febrile UTI and found a significantly higher pNGAL level than that in controls. The authors conclude that the use of rapid pNGAL assay at the time of admission can help predict the clinical course of febrile UTI in infants and determine treatment strategies. Urbschat et al. (2014) have reported a higher sNGAL level in adults with UTI accompanied by fever than in those without fever. A study of Petrovic et al. (2013) performed in 50 children with febrile UTI, demonstrates a significantly higher sNGAL level at admission than that at hospital discharge.

Conflicting data have been reported by Kim et al. (2014), who investigated 107 children with UTI and 337 children without UTI. In this large-scale pediatric study, sNGAL and uNGAL were elevated in the UTI compared to non-UTI group, but the difference between the two groups was statistically insignificant. When the comparison included the presence or absence of pyuria, the mean sNGAL and uNGAL content was significantly greater in the pyuria-posifive than that in pyuria-negative group. However, the optimum cut-off value 82.5 ng/ml for sNGAL (sensitivity 70%, specificity 60%, and AUC 0.67) and 10.3 ng/ml for uNGAL (sensitivity and specificity 70%, and AUC 0.78) demonstrated rather low sensitivity and specificity for diagnosing a pyuria-positive UTI.

Studies of Urbschat et al. (2014) and Kim et al. (2014) have demonstrated that uNGAL is more useful than sNGAL for the UTI diagnosis. In contrast, in the present study we found higher sensitivity and specificity of sNGAL than uNGAL for the febrile UTI diagnosis. Since a normal result of urinalysis does not enable the exclusion of UTI (Ghasemi et al. 2016; Salleeh et al. 2010; Etoubleau et al. 2009; Readorn et al. 2009), we conclude that sNGAL may be useful for the UTI diagnosis in febrile infants having no changes in urinalysis and for the differentiation of UTI from febrile states other than UTI. In most studies, urine and blood NGAL content was investigated in children with febrile UTI. We evaluated also infants with non-febrile UTI, because the UTI diagnosis and differentiation between contamination and infection is still a challenge at that especially vulnerable age. sNGAL was less sensitive for the non-febrile than febrile UTI, and uNGAL was unuseful for the non-febrile UTI diagnosis. In the present study, we also found a significantly higher level of systemic inflammatory markers in children with the febrile than non-febrile UTI, and in children with the non-febrile UTI compared to controls. We demonstrate that sNGAL associated with the level of WBC, CRP, ESR, and PCT, and uNGAL associated with the level of CRP and leukocyturia.

Although numerous studies have demonstrated an association between pNGAL, sNGAL, and uNGAL, on the one side, and WBC and CRP, on the other side (Ghasemi et al. 2016; Lee et al. 2015; Sim et al. 2015; Urbschat et al. 2014; Yim et al. 2014), other studies have found no such relationships (Seo et al. 2014; Hatipoglu et al. 2011; Ylmaz et al. 2009). Various studies have also demonstrated the association of sNGAL and uNGAL markers with leukocyturia and leukocyte esterase (Lee et al. 2015; Kim et al. 2014; Cullen et al. 2012; Decavele et al. 2011; Hatipoglu et al. 2011), but again some other studies reported no such relationships (Ylmaz et al. 2009). Seo et al. (2014) have speculated that NGAL content increases at the time of chronic kidney damage from bacterial infection, and decreases after clinical improvement. We opine that the association of sNGAL and uNGAL with fever and inflammatory markers, and a decrease in sNGAL and uNGAL after onset of treatment, support the notion that NGAL enhancement is induced by inflammatory process rather than by the kidney epithelial injury.

The limitations of our study include its design as a single-center study, a small number of patients, and a difference in children's age between the groups. However, it is challenging to have closely age-matched groups in observational, comparative studies. In this study, the age difference amounted to 2–3 months, which was rather small, despite being statistically significant. Therefore, it seems rather unlikely that a high content of serum and urine NGAL, we noted in children with febrile UTI, would be associated with the age-difference rather than with the severity of UTI.

In conclusion, this prospective, observational study demonstrates that serum NGAL is an excellent marker for the diagnosis of febrile UTI, having high sensitivity and specificity. Among the febrile infants, serum NGAL is a more sensitive marker for an early UTI diagnosis than urine NGAL. Among the non-febrile infants, serum NGAL is less sensitive for the UTI diagnosis and urine NGAL is not good to this end at all. In our opinion, serum NGAL may be a very useful marker for the UTI diagnosis in febrile infants, having normal results of urinalysis. Serum NGAL seems particularly useful for the differentiation between UTI and febrile states other than UTI-related and between infection and contamination in infants with a suspicion of UTI.

Conflicts of Interest The authors declare no conflicts of interest in relation to this article.

References

AAP (2011) American Academy of Pediatrics. Subcommittee on Urinary Tract Infection, Steering Committee on Quality Improvement and Management. Urinary Tract Infection: Clinical practice guideline for the diagnosis and management of the initial UTI in febrile infants and children 2 to 24 months. Pediatrics 128:595–610

Arambašić J, Mandić S, Debeliak Ž, Mandić D, Horvat V, Šerić V (2016) Differentiation of acute pyelonephritis from other febrile states in children using urinary neutrophil gelatinase-associated lipocalin (uNGAL). Clin Chem Lab Med 54(10):55–61

Bennett MR, Nehus E, Haffner C, Ma Q, Devarajan P (2015) Pediatric reference ranges for acute kidney injury biomarkers. Pediatr Nephrol 30:677–685

Cullen MR, Murray PT, Fitzgibbon MC (2012) Estabilishment of a reference interval for urinary neutrophil gelainase-associated lipocalin. Ann Clin Biochem 49:190–193

Decavele AS, Dhondt L, De Buyzere ML, Delanghe JR (2011) Increased urinary neutrophil gelainase associated lipocalin in urinary tract infections and leukocyturia. Clin Chem Lab Med 49(6):999–1003

Etoubleau C, Reveret M, Brouet D, Badier L, Brosset P, Fourcade L, Bahans C, Garnier F, Blanc P, Guigonis V (2009) Moving from bag to catheter for urine collection in non-toilet-trained children suspected of having urinary tract infection: a paired comparison of urine cultures. J Pediatr 154:803–806

Filho LT, Grande AJ, Colonetti T, Colonetti T, Della ÉSP, da Rosa MI (2017) Accuracy of neutropfil gelatinase-associated lipocalin for acute kidney injury diagnosis in children: systematic review and meta-analysis. Pediatr Nephrol. https://doi.org/10.1007/s00467-017-3704-6

Forster CS, Johnson K, Patel V, Wax R, Rodig N, Barasch J, Bachur R, Lee R (2017) Urinary NGAL deficiency in recurrent urinary tract infections. Pediatr Nephrol 32:1077–1080

Ghasemi K, Esteghamati M, Borzoo S, Parvaneh E, Borzoo S (2016) Predictive accuracy of urinary neutrophil gelatinase associated lipocalin (NGAL) for renal parenchymal involvement in children with acute pyelonephritis. Electron Physician 8:1911–1917

Hatipoglu S, Sevketoglu E, Gedikbasi A, Ylmaz A, Kiyak A, Mulazimoglu M, Aydogan G, Ozpacaci T (2011) Urinary MMP-9/NGAL complex in children with acute cystitis. Pediatr Nephrol 26:1263–1268

Ichino M, Kuroyanagi Y, Kusaka M, Mori T, Ishikawa K, Shiroki R, Kurahashi H, Hoshinaga K (2009) Increased urinary neutrophil gelatinase associated lipocalin levels in a rat model of upper urinary tract infection. J Urol 181(5):2326–2331

Kim BH, Yu N, Kim HR, Yun KW, Lim IS, Kim TH, Lee M-K (2014) Evaluation of the optimal neutrophil gelatinase-associated lipocalin value as a screening biomarker for urinary tract infections in children. Ann Lab Med 34:354–359

Kim BK, Yim HE, Yoo KH (2017) Plasma neutrophil gelatinase-associated lipocalin: a marker of acute pyelonephritis in children. Pediatr Nephrol 32:477–484

Krzemień G, Szmigielska A, Artemiuk I, Roszkowska-Blaim M (2014) False positive urine cultures in children under two years of age – own research. Dev Period Med 18(2):172–177

Lee H-E, Lee SH, Baek M, Choi H, Park K (2013) Urinary measurement of neutrophil gelatinase associated lipocalin and kidney injury molecule-1 helps diagnose acute pyelonephritis in a preclinical model. J Biomarkers 2013:6

Lee H-E, Kim DK, Kang HK, Park K (2015) The diagnosis of febrile urinary tract infection in children may be facilitated by urinary biomarkers. Pediatr Nephrol 30:123–130

McWilliam SJ, Antoine DJ, Sabbisetti V, Pearce RE, Jorgensen AL, Lin Y, Leeder JS, Bonventre JV, Smyth RL, Primohamed M (2014) Reference intervals for urinary renal injury biomarkers KIM-1 and NGAL in healthy children. Biomark Med 8:1189–1197

Nasioudis D, Witkin SS (2015) Neutrophil gelatinase-associated lipocalin and innate immune responses to bacterial infections. Med Microbiol Immunol 204:471–479

Nickavar A, Safaeian B, Valavi E, Moradpour F (2016) Validity of neutrophil gelatinase associated lipocaline as a biomarker for diagnosis of children with acute pyelonephritis. Pediatr Urol 13(5):2860–2863

Petrovic S, Bogavac-Stanojevic N, Peco-Antic A, Ivanisevic I, Kotur-Stevuljevic J, Paripovic D, Sopic M, Jelic-Ivanovic Z (2013) Clinical application neutrophil-associated lipocalin and kidney injury molecule-1 as indicators of inflammation persistence and acute kidney injury in children with urinary tract infection. Biomed Res Int 2013:947157

Rafiei A, Mohammadjafari H, Bazi S, Mirabi AM (2015) Urinary neutrophil gelatinase-associated lipocalin (NGAL) might be an independent marker for anticipating scar formation in children with acute pyelonephritis. J Renal Inj Prev 4:39–44

Readorn JM, Carstairs KL, Rudinsky SL, Simon LV, Riffenburgh RH, Tanen AD (2009) Urinalysis is not reliable to detect a urinary tract infection in febrile infants presenting to the ED. Am J Emerg Med 27:930–932

Salleeh H, McGillivray D, Martin M, Patel H (2010) Duration of fever affects the likelihood of positive bag urinalysis or catheter culture in young children. J Pediatr 156:629–633

Seo WH, Nam SW, Lee EH, Je B-K, Yim HE, Choi BM (2014) A rapid plasma neutrophil gelatinase-associted lipocalin assay for diagnosis of acute pyelonephritis in infants with acute febrile urinary tract infections: a preliminary study. Eur J Pediatr 173:229–232

Sim JH, Yim HE, Choi BM, Lee JH, Yoo KH (2015) Plasma neutrophil gelatinase-associated lipocalin predicts acute pyelonephritis in children with urinary tract infection. Pediatr Res 78:48–55

Singer E, Marko L, Paragas N, Barasch N, Dragun D, Müller DN, Budde K, Schmidt-Ott KM (2013) Neutrophil gelatinase-associated lipocalin: pathophysiology and clinical applications. Acta Physiol 207 (4):663–672

Stein R, Dogan HS, Hoebeke P, Kočvara R, Nijman RJM, Radmayr C, Tekgül S, European Association of Urology, European Society for Pediatric Urology (2015) Urinary tract infection in children: EAU/ESPU guidelines. Eur Urol 67:546–558

Szmigielska A, Krzemień G (2017) Sterile leukocyturia – difficult diagnostic problem in children. Dev Period Med 21(2):108–112

Tullus K (2011) Difficulties in diagnosing urinary tract infections in small children. Pediatr Nephrol 26:1923–1926

Urbschat A, Obermüller PP, Reissig M, Hadji P, Hofmann R, Geiger H, Gauer S (2014) Upper and lower urinary tract infections can be detected early but not be descriminated by urinary NGAL in adults. Int Urol Nephrol 46:2243–2249

Waikar SS, Sabbisetti VS, Bonventre JV (2010) Normalization of urinary biomarkers to creatinine during changes in glomerular filtration rate. Kidney Int 78 (5):486–496

Yim HE, Yim H, Bae ES, Woo SU, Yoo KH (2014) Predictive value of urinary and serum biomarkers in young children with febrile urinary tract infections. Pediatr Nephrol 29:2181–2189

Ylmaz A, Sevketoglu E, Gedikbasi A, Karyagar S, Kiyak A, Mulazimoglu M, Aydogan G, Ozpacaci T, Hatipoglu S (2009) Early prediction of urinary tract infection with urinary neutrophil gelatinase associated lipocalin. Pediatr Nephrol 24:2387–2392

Advs Exp. Medicine, Biology - Neuroscience and Respiration (2018) 38: 81–87
DOI 10.1007/5584_2017_123

Published online: 19 November 2017

A Two-Week Fecal Microbiota Transplantation Course in Pediatric Patients with Inflammatory Bowel Disease

Katarzyna Karolewska-Bochenek, Pawel Grzesiowski, Aleksandra Banaszkiewicz, Agnieszka Gawronska, Maria Kotowska, Marcin Dziekiewicz, Piotr Albrecht, Andrzej Radzikowski, and Izabella Lazowska-Przeorek

Abstract

Dysbiosis plays a major role in the etiology of inflammatory bowel disease (IBD). Fecal microbiota transplantation (FMT) is a new promising option for IBD treatment. We aimed to assess the effectiveness of a two-week FMT course in children with IBD. Ten patients, 10–17 years of age with moderate to severe IBD received a course of eight doses of freshly prepared FMT *via* a nasoduodenal tube or gastroscopy. All of the patients had pancolitis. There were eight cases of ulcerative colitis (UC) and two of Crohn's disease (CD). Disease activity was evaluated using the Pediatric UC Activity Index (PUCAI) and Pediatric CD Activity Index (PCDAI) for UC and CD, respectively, CRP, and fecal calprotectin on the day before the first infusion and then on the day before the next course of FMT. Clinical response, defined as a decrease of 15 points in either index, was observed in 9/10 patients (seven UC and two CD). Clinical remission, defined as a PCDAI score $\leq$ 10 and PUCAI score $<$ 10 measured at the same time point, was observed in 3/8 UC patients and 2/2 CD patients. Side effects observed were self-limiting and benign. We conclude that a short, intensive course of FMT has a beneficial effect on UC and CD colitis. FMT was well-tolerated and safe. Nonetheless, an optimal protocol of FMT administration is crucial for treatment efficacy.

Keywords

Children · Crohn's disease · Fecal microbiota transplantation · Inflammatory bowel disease · Ulcerative colitis

1 Introduction

Ulcerative colitis (UC) and Crohn's disease (CD) are chronic, progressive and devastating

K. Karolewska-Bochenek, A. Banaszkiewicz, A. Gawronska, M. Kotowska, M. Dziekiewicz, P. Albrecht, A. Radzikowski, and I. Lazowska-Przeorek (✉)
Department of Paediatric Gastroenterology and Nutrition, Warsaw Medical University, Warsaw, Poland
e-mail: ilazowska@gmail.com

P. Grzesiowski
Preventive Medicine and Rehabilitation Center, Infection Prevention Institute, Warsaw, Poland

diseases of the gastrointestinal tract (GI). The exact cause of these conditions remains unknown. The most current hypothesis states that in genetically susceptible individuals the damage of the GI tract stems from persistent inflammation driven by hyperactivation of the immune system (Kostic et al. 2014), which is caused by an altered gut microbiome (Kaakoush et al. 2012; Michail et al. 2012). Studies on the natural history of inflammatory bowel disease (IBD) show progression of the disease over time in both pediatric and adult populations (Benchimol et al. 2011; Pariente et al. 2011). However, IBD in children is usually more severe in the course and extension than in adults (Turner 2014). Therefore, the main aim of IBD management in children is to prevent progression of the disease, and to provide optimal physical development (Levine et al. 2014). Even now, in the era of biologic therapy, we still search for new powerful therapeutic tools. Because the role of the intestinal microbiota appears unquestionable in IBD (Campieri and Gionchetti 2001), the reinstatement of GI eubiosis could be a treatment target. This could be fulfilled in several ways: by antibiotic therapy, probiotic or prebiotic supplementation, and by fecal microbiota transplantation (FMT). Since the effectiveness of FMT in treating refractory *Clostridium difficile* infection (CDI) has been proven (van Nood et al. 2013; Borody et al. 2003) there have been experimental FMT therapies for several diseases, including IBD (Colman and Rubin 2014; Bennet and Brinkmann 1989). FMT therapy induces a remission in adults with UC in randomized trials (Moayyedi et al. 2015). One or two doses of FMT are curable for CDI, although serial FMT administration seems more efficacious in IBD therapy (van Nood et al. 2013). Data on the pediatric IBD population are very limited (Kellermayer et al. 2015; Suskind et al. 2015a, b; Kunde et al. 2013) and the exact scheme of FMT treatment has not yet been established. Therefore, the aim of our study was to assess the effectiveness of a two-week course of FMT in the treatment of pediatric patients with IBD.

2 Methods

This study was approved by the Clinical Research Ethics Committee of the Medical University of Warsaw, Poland. All the parents and children aged $\geq$16 years signed informed consent before participation in the study. All the patients and donors were informed of the benefits and potential risks of standardized FMT and laboratory screening.

2.1 Study Design and Population

This report is part of an ongoing, single-center, open, prospective trial that is being conducted at the Department of Pediatric Gastroenterology and Nutrition, at the Medical University of Warsaw, Poland, to assess the long-term effectiveness and safety of FMT in children with IBD. Our study was focused on the evaluation of the subjects who completed the initial two-week course of FMT.

Children and adolescents ranging from 10 to 17 years of age with UC or CD colitis refractory to standard therapy (e.g., steroids, immunomodulators, cyclosporine, tacrolimus, or anti-TNF agents) were asked to participate in the study. The diagnosis of CD and UC was based on clinical signs and symptoms, as well as on endoscopic, histological, and radiological results according to the Revised Porto criteria (Levine et al. 2014). The severity of CD and UC, and the subject's clinical response to therapy were evaluated using the Pediatric CD Activity Index (PCDAI) and Pediatric UC Activity Index (PUCAI). The participants began to be recruited in September 2014, and the study is still ongoing. The patients were at different disease stages, mild to severe UC (PUCAI: 10–70 points) and mild to moderate CD colitis (PCDAI: 10–37.5 points), and underwent different treatment processes.

We also assessed the markers of inflammation: CRP and fecal calprotectin.

2.2 Donor Selection

Donors were screened for human immunodeficiency virus (HIV), hepatitis A virus (HAV-Ab) (IgM and IgG), hepatitis B virus (HBV) (HBsAg, anti-HBc), HCV (Ig G anti HCV), Epstein-Barr virus (EBV) (IgM and IgG), cytomegalovirus (CMV) (IgM and IgG), *Treponema pallidum*, and underwent stool testing for *Salmonella*, *Shigella*, *Campylobacter*, *Yersinia*, *Escherichia coli*, *Clostridium difficile* toxins, pathogenic flora, ova, and parasites. Donors were generally healthy, on a regular diet, and had no antibiotic treatment during the last 3 months before donating the material. They were not related to patients.

2.3 Stool Preparation

The stools were collected at home. The donors produced a single stool sample up to 4 h before FMT preparation. The stool samples were placed in sterile tubes, immediately frozen in −20 °C, and forwarded to the Department of Microbiology Laboratory. Donated stools with a weight of approximately 50 g were thawed and mixed with 30 ml sterile saline to prepare a homogeneous suspension. Next, the suspension was filtered several times through suitable filter under biological safety cabinet class 2 conditions. The residue on the filter was mixed with 30 ml of sterile saline and was slowly filtered under the same conditions. The final volume of approximately 50 ml was placed in sterile syringe ready for use. The transfer was given to the recipient immediate after preparation.

2.4 Study Intervention

At screening, the demographic data, medical history, and physical examination were obtained. All the patients with IBD underwent disease activity assessment using the PUCAI for UC and PCDAI for CD at study onset and after 18–33 days of the last FMT administration. Previous and current treatment information was recorded. The subjects were maintained on their pre-transplant medications. They did not receive any antibiotics or bowel lavage before transplantation. Prior to FMT therapy, they received a proton pump inhibitor at a dose 1 mg/kg *per os* and the anti-vomiting drug ondansetron at a dose of 4–8 mg *per* os in the morning on the day of FMT administration.

Each patient received 50 ml of FMT *via* gastroscopy or a nasoduodenal tube on 5 consecutive days during the first week and every other day during the second week (in total, 8 FMT doses in 12 days). FMT was administered slowly in 10-min sessions. The nasoduodenal tube or gastroscope was then flushed with 10 ml of normal saline. During each procedure, the patients were seated in an upright position. All the patients were monitored for 4 h and were allowed to drink 1 h and eat 2 h after FMT administration. All the gastroscopies were performed under general anaesthesia. The scheme of study course is presented in Fig. 1.

2.5 Outcomes

The primary outcome measure was a clinical response, defined as a decrease of 15 points or more in PUCAI or PCDAI measured 18–33 days after the last dose of FMT. The secondary outcome measures were as follows: (1) clinical remission defined as a PCDAI score decrease ≤10 for CD and a PUCAI score decrease <10 for UC measured at the same time point; (2) decrease in calprotectin; and (3) evidence of adverse effects measured for 4 h after the procedure.

3 Results

A total of 10 (3 boys) subjects participated in the study including 8 patients with UC and 2 patients with CD colitis. The basic characteristics of the study participants are presented in Table 1.

Clinical response was observed in nine out of the ten patients (7 UC and 2 CD). Clinical

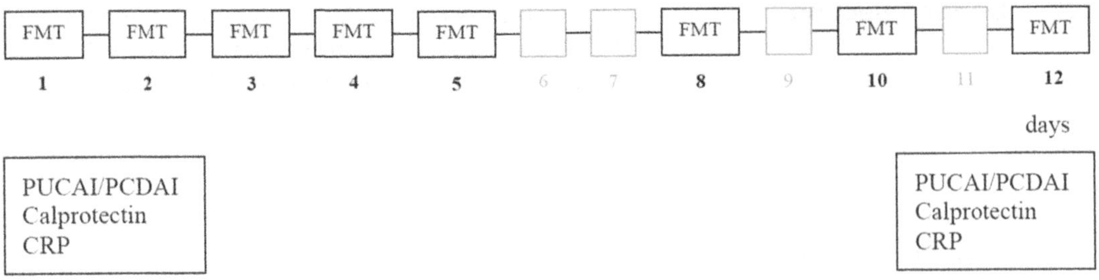

Fig. 1 Scheme of the study course. *FMT*, fecal microbiota transplantation; *PCDAI*, pediatric Crohn's disease activity index; *PUCAI*, pediatric ulcerative colitis activity index

Table 1 Basic characteristics of fecal microbiota transplantation (FMT) patients

Patient no.	Sex	Age (year)	Disease duration (year)	Diagnosis	Paris classification	Treatment
1	F	17	13	UC	E4 S0	GCS, AZA, 5ASA, cycA, Tacro
2	F	17	10	UC	E3 S0	GCS, AZA, 5ASA, cycA, Tacro
3	F	17	4	UC	E4 S1	GCS, AZA, 5ASA, cycA, Tacro, IFX, ADA
4	M	10	6	UC	E4 S1	GCS, AZA, 5ASA, cycA, Tacro
5	M	12	2	UC	E4 S1	GCS, AZA, 5ASA
6	F	16	2	UC	E4 S1	GCS, AZA, 5ASA
7	F	16	4	UC	E S0	GCS, AZA, 5ASA
8	F	12	3.5	CD	A1b L2 B1	GCS, AZA, 5ASA, IFX, MTX
9	M	17	2	CD	A2 L2 B1	GCS, AZA, 5ASA
10	F	17	1	UC	E4 S0	GCS, AZA, 5ASA

UC ulcerative colitis, *CD* Crohn's disease, *ADA* adalimumab, *5ASA* aminosalicylates, *AZA* azathioprine, *cycA* cyclosporine A, *IFX* infliximab, *GCS* glucocorticosteroids, *Tacro* tacrolimus

remission was achieved in 3 out of the 8 patients with UC and in all (2/2) patients with CD. Calprotectin was decreased in 5 out of the 10 patients. Unfortunately, two out of the ten patients did not have control results. Two of the eight patients (number 4 and 7) had clinical remission and a decrease in CRP and calprotectin levels. Detailed changes in disease activity are shown in Table 2.

3.1 Safety and Adverse Events

A two-week FMT course was very safe. We observed only mild, self-limiting adverse effects such as vomiting in two patients after the first dose of FMT. Additionally, one patient presented with nausea. The FMT procedure was tolerated in all the patients except three episodes of vomiting within 2 h after infusion.

4 Discussion

The results of this prospective study show that a two-week course of FMT was clinically effective in most children and adolescents with IBD. For UC patients, clinical response and clinical remission were achieved in seven and three out of the eight patients, respectively. All (2/2) patients with CD achieved clinical remission.

The data regarding the assessment of FMT in children with IBD are limited. Until now, 26 patients in four observational studies have been assessed. Our findings are in line with the results of two of those studies that assessed the effectiveness of multiple doses of FMT in

Table 2 Patients' laboratory data before and after the course of fecal microbiota transplantation (FMT)

Patient no.	Disease activity		CRP		Calprotectin	
	Before FMT PUCAI/PCDAI	After FMT PUCAI/PCDAI	Before FMT (mg/dl)	After FMT (mg/dl)	Before FMT (μg/g)	After FMT (μg/g)
1	60	15	0.4	0.5	>1,800	1,502
2	70	40	0,7	0.6	156	Not done
3	50	0	3.7	2.6	824	1,562
4	15	0	0.6	0.1	>1,800	1,583
5	40	10	1.2	0.1	>1,800	632
6	50	20	0.6	1.2	>1,800	>1,800
7	35	0	0.4	0.1	>1,800	<100
8	37.5	7.5	0.9	0.8	>1,800	>1,800
9	15	0	0.4	0.4	>1,800	Not done
10	25	15	0.7	0.1	>1,800	1,786

CRP C-reactive ptotein, *FTM* fecal microbiota transplantation, *PCDAI* pediatric Crohn's disease activity index, *PUCAI* pediatric ulcerative colitis activity index

pediatric UC. In Kunde et al.'s (2013) study, seven out of the nine children showed clinical response after five FMT enemas which were administered in five consecutive days. In Kellemayer et al.'s (2015) study full remission was achieved in all three UC patients who received multiple doses of FMT. However, a single-dose of FMT *via* nasogastric tube in active pediatric UC turned to be ineffective (Suskind et al. 2015b). We found a very good response in CD patients (2/2), which is in line with the only previous study that evaluated FMT in children with CD (Suskind et al. 2015a).

A direct comparison between our results and the above-mentioned studies is very difficult for a few reasons. Firstly, a different route of FMT administration, *via* a nasogastric tube (Suskind et al. 2015a; b) or *via* enema (Kellermayer et al. 2015; Kunde et al. 2013) was used in previous studies. We decided to administer FMT directly to the descending part of the duodenum. With a nasogastric tube, there is an increased risk of destruction or weakening of a part of the fecal microbiota due to contact with the gastric contents. FMT administration *via* enema, on the other hand, may not enable fecal microbiota to reach the entire colon. Secondly, different FMT quantity and volumes or different intervals between FMT were given. Although the optimal FMT scheme for IBD treatment has not yet been established, it seems that multiple dose of FMT are indicated in UC management, while even one dose of FMT may be sufficient to achieve a favorable response or even remission in CD. Thirdly, different premedication before FMT was used. In the present study, and that of Kunde et al.'s (2013), the subjects did not receive any antibiotic pre-treatment or bowel preparation. By contrast, in the Suskind et al.'s (2015a) study, the recipients received rifaximin and macrogol. Randomized controlled trials are needed to establish the most appropriate protocol of premedication before FMT. Finally, the patients described in those studies are heterogeneous; they had different disease durations and they were on various IBD therapies. Notwithstanding these differences, we would like to emphasize that the two-week course of FMT was effective in treating IBD in children. A major limitation of our study is the assessment of an early response. However, the study is still going on and we hope we will report durable effects of FMT therapy in the future.

In the present study, we demonstrate a decrease in calprotectin content in five out of the ten 10 patients; a finding that is similar to that of Suskind et al. (2015b), who a decline in calprotectin in three out of the four patients during a 12-week-long follow-up. These findings are similar, even though we assessed the calprotectin content after the 8th FMT infusion as opposed to the effect of a single infusion in the study quoted above. There are no other pediatric or adult papers on the subject, which makes a broader comparison impossible. Our present findings suggest that FMT not only ameliorates IBD symptoms but also is conducive to mucosal healing, an important sign of treatment efficacy and a prognostic marker in IBD.

FMT in patients with IBD is safe and tolerable. Side effects seem closely related to the route of administration. Our data are in line with those from previous surveys conducted in the pediatric population (Kellermayer et al. 2015; Suskind et al. 2015a, b; Kunde et al. 2013). We observed only a few side effects, mainly emanating from the gastrointestinal tract, and thee effects short-term, self-limiting, and tolerable. The most common side effect was vomiting or nausea, occurring up to 2 h after suspension administration. We presume that these effects might relate to the speed of FMT infusion. Prolonging the infusion time from 5 to 10 min improved the tolerance to the procedure. The side effects we observed are akin to those observed after FMT administration through a nasogastric or naso-jejnunal route (Kellermayer et al. 2015; Rossen et al. 2015; Suskind et al. 2015a, b). Nonetheless, more severe side effects, such as small bowel abscess or aspiration pneumonia, have been reported in adult IBD patients (Moayyedi et al. 2015; Rossen et al. 2015).

The main advantage of the present study was the assessment of a new, multiple-dose intensive scheme of FMT therapy in children and adolescents with IBD. The data regarding FMT treatment of pediatric IBD patients are limited, but it appears that multiple doses of FMT are more efficacious than a single-dose scheme in treating UC (Kellermayer et al. 2015; Suskind et al. 2015b; Kunde et al. 2013). By contrast, a single dose of FMT may be sufficient for CD patients. Importantly, we chose a route of FMT administration *via* a descending part of the duodenum, likely the best one as it reduces the gastric acid effect on the fecal microbiota, making it more viable. Although our study had a small sample size, it is one of the largest that has been published yet, concerning the pediatric population. We also confirmed the safety and good tolerability of FMT in children with IBD.

Our study has several limitations. It was an observational study; thus, we could not assess the real efficacy of fecal microbiota transplantation. The study group was not homogeneous as the patients were of different age and showed different disease activity. They also received additional treatments, although these treatments remained unchanged throughout the study time. Despite these limitations, we believe we have added to the evidence of the effectiveness of fecal microbiota transplantation for the treatment of inflammatory bowel disease.

5 Conclusions

This study provides data that an intensive two-week course of fecal microbiota transplantation is effective as a new treatment option in children with the inflammatory bowel disease that is refractory to conventional therapies. However, the index of clinical remission in the short-term was rather poor. Nonetheless, FMT is safe and well-tolerated in pediatric patients suffering from ulcerative colitis and Crohn's disease.

Conflicts of Interest The authors declare that they have no competing interests in relation to this article.

References

Benchimol EI, Fortinsky KJ, Gozdyra P, Van den Heuvel M, Van Limbergen J, Griffiths AM (2011) Epidemiology of pediatric inflammatory bowel disease: a systematic review of international trends. Inflamm Bowel Dis 17(1):423–439

Bennet JD, Brinkmann M (1989) Treatment of ulcerative colitis by implantation of normal colonic flora. Lancet 1(8630):164

Borody TJ, Warren EF, Leis S, Surace R, Ashman O (2003) Treatment of ulcerative colitis using fecal bacteriotherapy. J Clin Gastroenterol 37(1):42–47

Campieri M, Gionchetti P (2001) Bacteria as the cause of ulcerative colitis. Gut 48(1):132–135

Colman RJ, Rubin DT (2014) Fecal microbiota transplantation as therapy for inflammatory bowel diseases: a systematic review and meta-analysis. J Crohns Colitis 8(12):1569–1568

Kaakoush NO, Day AS, Huinao KD, Leach ST, Lemberg DA, Dowd SE, Mitchell HM (2012) Microbial dysbiosis in pediatric patients with Crohn's disease. J Clin Microbiol 50(10):3258–3266

Kellermayer R, Nagy-Szakal D, Harris RA, Luna RA, Pitashny M, Schady D, Mir SA, Lopez ME, Gilger MA, Belmont J, Hollister EB, Versalovic J (2015) Serial fecal microbiota transplantation alters mucosal gene expression in pediatric ulcerative colitis. Am J Gastroenterol 110(4):604–606

Kostic AD, Xavier RJ, Gevers D (2014) The microbiome in inflammatory bowel diseases: current status and the future ahead. Gastroenterology 146(6):1489–1499

Kunde S, Pham A, Bonczyk S, Crumb T, Duba M, Conrad H Jr, Cloney D, Kugathasan S (2013) Safety, tolerability and clinical response after fecal microbiota transplantation in children and young adults with ulcerative colitis. J Pediatr Gastroenterol Nutr 56 (6):597–601

Levine A, Koletzko S, Turner D, Escher JC, Cucchiara S, de Ridder L, Kolho KL, Veres G, Russell RK, Paerregaard A, Buderus S, Greer ML, Dias JA, Veereman-Wauters G, Lionetti P, Sladek M, Martin de Carpi J, Staiano A, Ruemmele FM, Wilson DC, ESPGHAN (2014) ESPGHAN revised Porto criteria for the diagnosis of inflammatory bowel disease in children and adolescents. J Pediatr Gastroenterol Nutr 58(6):795–806

Michail S, Durbin M, Turner D, Griffiths AM, Mack DR, Hyams J, Leleiko N, Kenche H, Stolfi A, Wine E (2012) Alterations in the gut microbiome of children with severe ulcerative colitis. Inflamm Bowel Dis 18 (10):1799–1808

Moayyedi P, Surette MG, Kim PT, Libertucci J, Wolfe M, Onischi C, Armstrong D, Marshall JK, Kassam Z, Reinisch W, Lee CH (2015) Fecal microbiota transplantation induces remission in patients with active ulcerative colitis in a randomized, controlled trial. Gastroenterology 149(1):102–109

Pariente B, Cosnes J, Danese S, Sandborn WJ, Lewin M, Fletcher JG, Chowers Y, D'Haens G, Feagan BG, Hibi T, Hommes DW, Irvine EJ, Kamm MA, Loftus EV Jr, Louis E, Michetti P, Munkholm P, Oresland T, Panés J, Peyrin-Biroulet L, Reinisch W, Sands BE, Schoelmerich J, Schreiber S, Tilg H, Travis S, van Assche G, Vecchi M, Mary JY, Colombel JF, Lémann M (2011) Development of the Crohn's disease digestive damage score, the Lémann score. Inflamm Bowel Dis 17(6):1415–1422

Rossen NG, Fuentes S, van der Spek MJ, Tijssen JG, Hartman JH, Duflou A, Löwenberg M, van den Brink GR, Mathus-Vliegen EM, de Vos WM, Zoetendal EG, D'Haens GR, Ponsioen CY (2015) Findings from a randomized controlled trial of fecal transplantation for patients with ulcerative colitis. Gastroenterology 149(1):110–118

Suskind DL, Brittnacher MJ, Wahbeh G, Shaffer ML, Hayden HS, Qin X, Singh N, Damman CJ, Hager KR, Nielson H, Miller SI (2015a) Fecal microbial transplant effect on clinical outcomes and fecal microbiome in active Crohn's disease. Inflamm Bowel Dis 21(3):556–563

Suskind DL, Sangh N, Nielson H, Wahbeh G (2015b) Fecal microbiota transplantation via nasogastric tube for active pediatric ulcerative colitis. J Pediatr Gastroenterol Nutr 60(1):27–29

Turner D (2014) Relapsing and refractory ulcerative colitis in children. Dig Dis 32(4):419–426

van Nood E, Vrieze A, Nieuwdorp M, Fuentes S, Zoetendal EG, de Vos WM, Visser CE, Kuijper EJ, Bartelsman JF, Tijssen JG, Speelman P, Dijkgraaf MG, Keller JJ (2013) Duodenal infusion of donor feces for recurrent *Clostridium difficile*. N Engl J Med 386(5):407–415

Advs Exp. Medicine, Biology - Neuroscience and Respiration (2018) 38: 89–98
DOI 10.1007/5584_2017_108

Published online: 5 October 2017

Specialized Pain Clinics in Primary Care: Common Diagnoses, Referral Patterns and Clinical Outcomes – Novel Pain Management Model

Yacov Fogelman, Eli Carmeli, Amir Minerbi, Baruch Harash, and Simon Vulfsons

Abstract

An estimated 19% of the adult population in western countries lives with chronic pain. Pain management lies mainly within the primary care and community setting. We evaluated the outcome of a new model of secondary care clinics, conducted by primary care physicians with specialized training in pain medicine. Data on referral patterns, prevalence of pain diagnosis, and medication consumption were recorded at five secondary pain management clinics in the community setting. In total, 997 patients with pain attended 2,652 visits (average 2.7 visits *per* patient) during 12 mo. Patients' age ranged from 18 to 92 yr (mean 59 ± 19). Mean pain intensity on the first visit, evaluated by the visual analogue scale was 7.7/10. Myofascial pain syndrome was the most common diagnosis (82%). Treatment included dry needling or trigger point injection (82%), manual myofascial release (23%), and pharmacotherapy (38%). Significant short-term improvement after treatment was reported by 75% of patients, and 72% reported long-term improvement. Four percent were referred to tertiary care pain clinics, 5% were referred to other specialists, and 5% to imaging. Secondary, community-based pain clinics, run by specially-trained primary physicians, demonstrated feasibility. The vast majority of patients referred to the clinics were treated using simple, inexpensive modalities, while sparing referrals to unnecessary consultation visits, imaging tests, and medications.

Keywords

Myofascial pain syndrome · Pain · Primary care physician · Secondary pain management

Y. Fogelman (✉), A. Minerbi, B. Harash, and S. Vulfsons
Institute of Pain Medicine, Rambam Health Care Campus, Haifa, Israel

The Ruth and Bruce Rappaport Faculty of Medicine, Technion – Israel Institute of Technology, 8 Efron Street, Haifa, Israel
e-mail: yfogelman@gmail.com

E. Carmeli
Department of Physical Therapy, University of Haifa, 199 Aba Khoushy Ave., Haifa, Israel

1 Introduction

Chronic pain has been described as a world health problem, and its management is a fundamental human right (Declaration of Montréal 2017; Minerbi and Vulfsons 2013). The prevalence of moderate-severe chronic pain (more than 6 mo) is 19% among adult Europeans (Turk et al. 2016). The long-term effects of chronic pain can be far reaching and include neurobiological and psychosocial consequences. Work productivity, quality of life, self-efficacy, coping, mood, sleep and appetite disorders, depression, fear, and anxiety are all negatively influenced by chronic pain (Mehta et al. 2016; Kroenke 2014; Breivik et al. 2006). Chronic pain also has a detrimental effect on patents' families and caregivers, who often become subject to great physical and psychological burdens in their efforts to be supportive (Dueñas et al. 2016; Carmeli 2014).

Due to a large number of patients and the paucity of tertiary pain care management facilities, pain is undertreated. Waiting lists for pain specialists can be extremely long, severely compromising professional pain consultation (Thomsen et al. 2002). While the care of chronic pain patients remains predominantly in the realm of primary care physicians (PCPs), PCPs with adequate knowledge of managing these patients are in short supply (Declaration of Montréal 2017; Peppin et al. 2015; Lincoln et al. 2013; Leverence et al. 2011). As a result, pain management services in the community lead to suboptimal and fragmented pain care, which has a negative impact on both individual patient care and public health.

The suboptimal management of chronic pain calls for a paradigm change. One proposal for such change (Minerbi and Vulfsons 2013) suggests the development of a pain service based on three levels of care: (1) PCPs; (2) secondary care clinics, operated primarily by PCPs trained in pain management; and (3) pain specialists working in tertiary care centers. According to this model, a school for pain medicine was established in Israel in 2010. The aim was the provision of in-depth year-long training for PCPs in the management of chronic pain. Since initiation of the program, six pain management education courses for PCPs have been completed at the Technion School for Continuing Medical Education. One hundred and thirty nine physicians who completed the first course received a certificate of 'Pain Trustee', testifying to 120 h of theoretical and practical education in pain medicine (50% theory and frontal lectures, 50% hands-on teaching). Fifty-six of these physicians participated in an additional course of 120 h in total (25% theoretical and 75% hands-on clinical teaching), together with a practicum of an additional 60 h. Completion of the two courses and practicum, totaling 300 h, entitled the participants to a diploma in Pain and Musculoskeletal Medicine. Five physicians with this diploma also received practical training at the Rambam Institute for Pain Medicine; a large tertiary center pain clinic based in Haifa, Israel. They dedicate part of their clinical time as family physicians to pain management. As 'secondary pain trustees', they treat their own patients with pain, and patients referred by other PCPs.

The major aim of the present study was to describe the activity in these five secondary pain clinics in terms of diagnoses made, treatments given, measures of clinical outcomes, imaging studies conducted, and further referral. As such, this study examined the feasibility of a model of pain management run by specialized family physicians.

2 Methods

The study received approval of the institutional ethics committee of both Clalit Health Services and the institutional review board of the University of Haifa. Data collected for the purpose of the study were saved in a coded manner, to protect the patients' privacy. This observational study was conducted in five community-based secondary pain clinics.

2.1 Patients and Participating Clinics

The patients who participated in this study either belonged to the practices of the primary physicians who conducted the study, or were referred for consultation by fellow family physicians. All participants' visits in the five clinics were documented in a digital medical records system. Eligibility criteria for inclusion in the study were the presentation of pain as the main complaint, with at least one-week duration. Patients underwent full history taking and physical examination. After a diagnosis was made according to physicians' judgment using the international classification of diseases (ICD)-10 version, they were offered a treatment series in the clinic or were referred to a tertiary pain clinic at a nearby hospital. Data were collected during clinic visits and entered by physicians into a dedicated patient registry.

This multi-center study was conducted in five community-based, secondary pain clinics, managed by specialized family physicians who had undergone extensive pain management training. Three clinics were situated in major cities, one in a small town, and one in a rural setting. The clinics belong to three of the four health medical organizations in Israel (Clalit, Leumit, and Meuchedet).

2.2 Data Collected

Consecutive available patients who met the inclusion criteria were invited to enroll in the study. For each patient who agreed to participate, the following information was recorded by the physicians: date of birth, gender, generic quality of life (e.g., general health, physical/mental/social role functioning), pain location, severity, and duration. At each visit, the following were recorded: the medical diagnosis reached, the treatment given, and the referral for further workup when deemed necessary imaging, expert consultation (such as rheumatologist, orthopedic surgeon, and neurologist). On follow-up visits, self-reported changes in pain were recorded (visual analogue scale of 1–10), including pain intensity, duration, and functional limitations. Self-reported improvement in pain was recorded on a 0–5 point likert scale.

2.3 Physicians Participating in the Study

Five family physicians working as secondary pain management practitioners in five community health care clinics participated in the study. These primary care physicians had undergone special training and are certified pain trustees with a diploma in pain and musculoskeletal medicine (see above). Data collected included demographic characteristics; location, duration, and mean intensity of pain on a 10-level visual analogue scale; patients' diagnoses, treatment modality, number of treatments and improvement after treatment (as assessed on follow-up visits); and referrals for imaging or further consultation.

Most patients in Israel have adequate access to their primary care physician. The patients in our sample were referred by other PCPs (predominantly) or by other primary caregivers such as orthopedic surgeons, neurologists and rheumatologists.

2.4 Statistical Analysis

Quantitative results were reported as means ± SD. All analyses were carried out with the SPSS v12 statistical package.

3 Results

We documented 997 pain patients who attended the five clinics included in this study. We recorded a total of 2,652 visits for pain management (average 2.7 ± 2.8 visits *per* patient) during 12 mo. Patients' age ranged from 18 to 92 yr (mean 59 ± 19). Of the 997 patients, 618 (62%) were female. The mean pain score before treatment was 7.7 ± 1.3, on a 10-point visual analog scale. Table 1 presents the duration of pain reported by the patients. Most of them reported chronic pain, over 6 mo in duration (Table 1). Forty percent complained of low back pain, 17% of cervical/head pain, and 14% of shoulder pain. Table 2 presents the prevalence of clinical diagnoses reached. Myofascial pain syndrome (MPS) was by far the most prevalent diagnosis (82%), followed by radicular pain (7%), osteoarthritis (6%), and chronic widespread pain (6%). Table 3 summarizes the pain management modalities applied. Dry needling and trigger point injection were the dominant modes of therapy.

Figure 1 presents the short term (up to one month) improvement in pain after treatment. Seventy-five percent of patients reported significant improvement (more than 70%) or complete pain relief. Telephone follow-up was performed at 6 mo to all the patients. Figure 2 presents the long term (six month) improvement after treatment. Seventy-two percent of the patients reported significant or complete pain relief. In Fig. 3 we compared the self-reported short-term improvement after treatment in acute pain *vs.* chronic pain patients. Patients with acute pain were more likely to report significant and complete recovery than chronic patients.

Referral for further imaging studies was relatively scarce, with only 6% of participants referred to further studies. Likewise, only 5% of the participants were referred for further specialist consultation. Four percent were referred to a tertiary care pain clinic in a nearby hospital for reevaluation, multidisciplinary treatment, or to undergo invasive procedures.

Table 1 Duration of patients' pain (n = 906).

Pain duration	%
1 wk to 6 mo	37%
7–18 mo	27%
19–59 mo	20%
5–10 yr	7%
11–20 yr	6%
>20 yr	3%

4 Discussion

The evolution of pain clinics was driven by the vast demand for their services. This demand is based mainly on the enormous numbers of patients that experience pain and the paucity of adequate solutions in health systems world-wide. In Israel, the majority of pain management services are under the auspices of tertiary care clinics of large hospitals. The establishment of such centers has resulted in rising expectations among patients for treatment and it is a driving force for a change in the way pain management services are organized and commissioned. Although chronic pain treatment is well established, the discrepancy between demand and supply is huge. In Europe, 70% of patients with chronic pain consults their family

Table 2 Clinical diagnostic characteristics of patients (n = 997).

Diagnosis	%
Myofascial pain syndrome (MFS)	82%
Radicular pain	7%
Chronic widespread pain	7%
Osteoarthritis	5%
Migraine	4%
Systemic inflammatory diseases (rheumatoid arthritis, vasculitis)	3%
Nerve entrapment (such as carpal tunnel syndrome)	1%
Other (including cancer pain, post herpetic neuralgia, post-operative pain, complex regional pain syndrome)	6%

Table 3 Pain management modality applied (n = 972).

Treatment	%
Dry needling	82%
Pharmacotherapy	38%
Manual myofascial release	23%
Kinesiotaping	7%
Prolotherapy	0.5%

physicians, 28% with orthopedic surgeons, and a mere 2% are managed by pain physicians (Breivik et al. 2006).

Problems in the management of patients with chronic pain stem from three major issues: (1) high prevalence of chronic pain; (2) lack of knowledge in the management of chronic pain by primary care physicians; (3) poor availability of

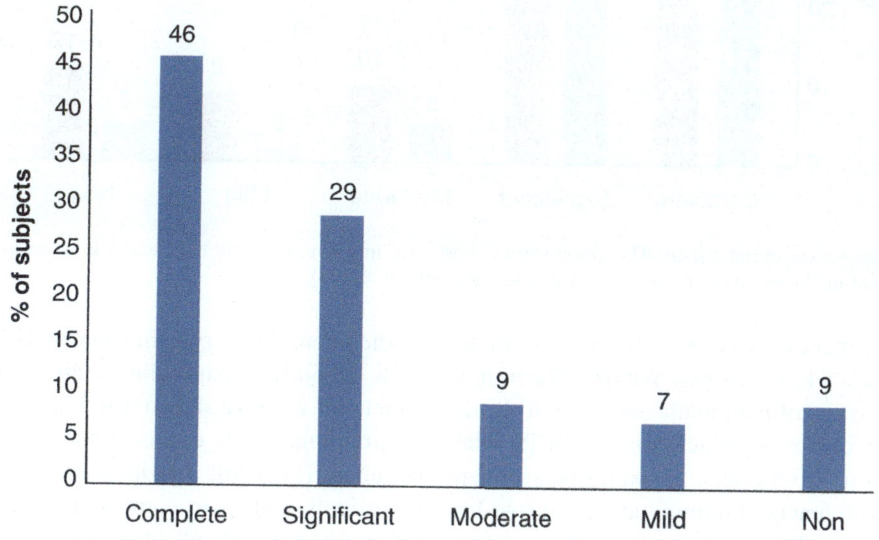

Fig. 1 Short-term, up to one month, improvement in pain perception after treatment. Data are presented for all patients (n = 796).

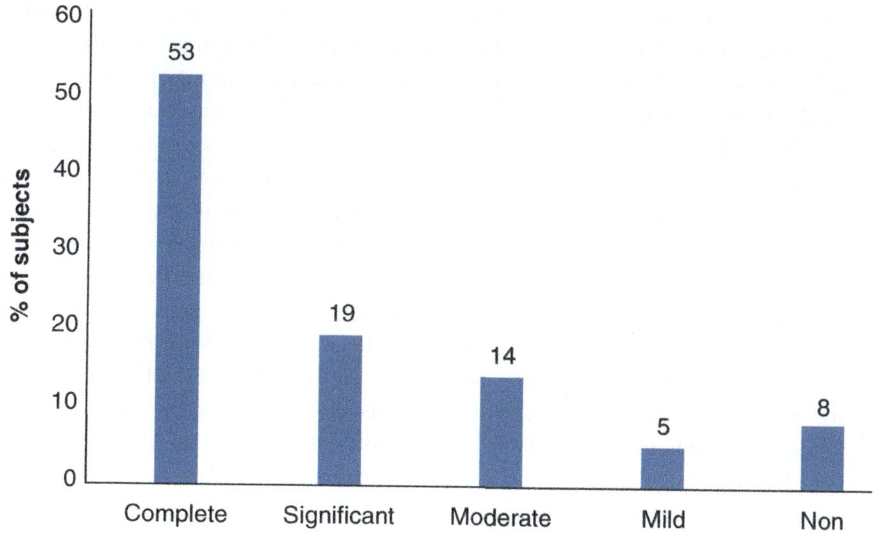

Fig. 2 Long-term, more than six month, improvement in pain perception after treatment. Data are presented for patients from three clinics (n = 466).

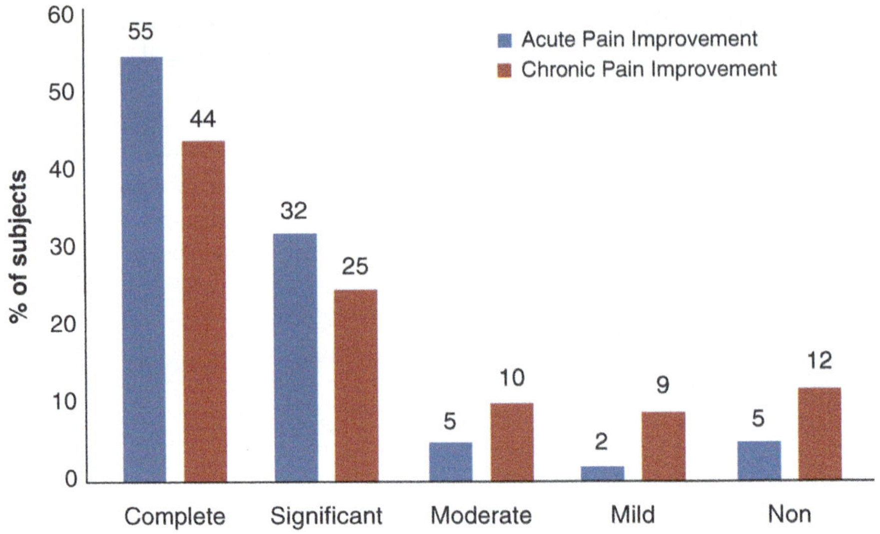

Fig. 3 Acute *versus* chronic pain. The graph shows short-term improvements after treatment in acute pain patients (n = 238) and in chronic pain (more than six months) patients (n = 507).

pain consultancy services. Resolving these challenges is a long process. We describe in this paper outcomes of pain management in five primary care clinics in Israel, run by PCPs who completed a 120-h course in pain management and received practical training at an institute for pain medicine. This study demonstrates the feasibility of training primary care physicians to provide care to 95% of pain patients they see in their daily practice, albeit with a significant educational effort. The vast majority of these patients have musculoskeletal pain, a condition quite neglected in medical schools, family practice, and orthopedic residency programs (Matzkin et al. 2005; Freedman and Bernstein 2002). The workload of PCPs involves the treatment and management of acute and chronic (Gureje et al. 2001). Pain complaints, can comprise up to 40% of the main causes for patients' visits to their family doctors (Mäntyselkä et al. 2001). Thus, primary and secondary pain management services are a favorable solution for managing patients with pain, especially chronic pain. All fields of medicine benefit from community-based medical experts. We believe that pain medicine can also benefit from certified physicians serving as secondary referral addresses, mid-way between primary care medicine and tertiary pain centers. To date, few such secondary pain clinics are available in Israel, but we have shown that with the appropriate training, PCPs can continue to serve their communities, while handling the vast majority of patients with pain complaints. We represent a flow chart of patients in secondary pain clinics in Fig. 4.

To our knowledge, this is the first study to report the results of a model of care focusing on secondary pain clinics in primary care, characterized by trained physicians whose primary specialty is family practice with additional competence in musculoskeletal and pain medicine. This study highlights the important role of secondary pain clinics in promoting optimal care for persons with pain. Such clinics offer care at the forefront of their patients' interface with the healthcare system and are particularly well positioned to identify gaps and strengths of the healthcare system, with its attempts to improve pain care. Other models exist, with varying degrees of emphasis on primary care physician empowerment in knowledge and treatment skills (Speerin et al. 2014). In such models, additional elements are also found such as patient education, web-based information and community-based group therapy. Our model was aimed and

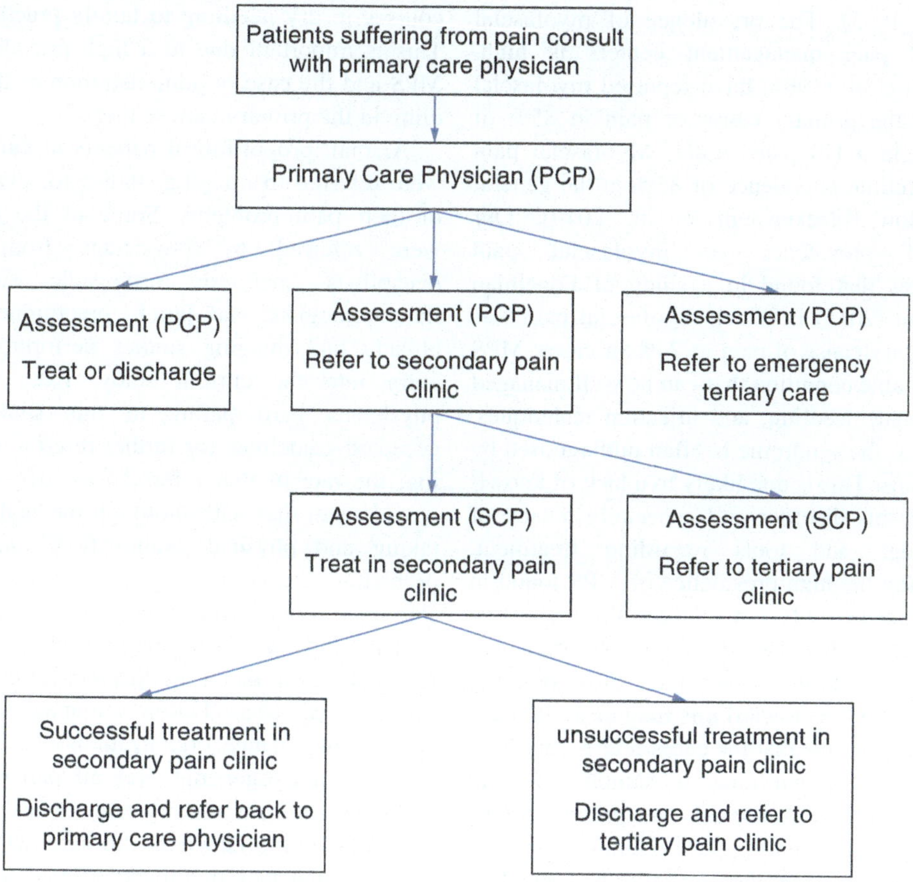

Fig. 4 Flow-chart for patients in secondary pain clinics.

targeted at the deficiency of knowledge and skill that primary care physicians have, and in reducing that gap. Although the training is quite intense (300 h), the rewards appear to be clear.

The Veterans Health Administration has introduced a national pain care strategy to meet the needs of veterans, called Stepped Care Model for Pain Management (VHA Directive 2009-053). The program emphasizes an individualized approach to pain management. This model provides diagnosis and treatment of pain patients within primary care. As medical problems increase in complexity or patients fail to achieve treatment goals with more conservative interventions, they are referred to specialized care and interdisciplinary approaches. The advantages of this model include greater accessibility to pain management, fewer referrals, reductions in treatment costs and empowering of the primary physician. This model, which has some resemblance to our model, is advocated by the American Academy of Pain Medicine (Dubois et al. 2009), and it has been cited by the Institute of Medicine as a potentially important model of care for pain management (Steglitz et al. 2012).

Of our patients included in the current work, 82% were diagnosed with myofascial pain syndrome (MPS). Myofascial pain is one of the most common sources of pain (Skootsky et al. 1989) and it is a major reason of disability and dysfunction (WHO 2003). Patients present with symptoms such as back pain, headache, limb pain, and neck and shoulder pain. If left untreated, myofascial pain can lead to chronic pain syndromes (Kalichman and Vulfsons 2010;

Gerwin 1995). The prevalence of myofascial pain in pain management centers is high. Fishbain et al. (1986) have reported myofascial pain as the primary cause of pain in 85% of patients in a US pain center. Myofascial pain has a lifetime prevalence of 85% in the general population (Fleckenstein et al. 2010). Our reported prevalence of myofascial pain resembles that found in a study of Fogelman and Kent (2015) in which myofascial pain was the primary cause of pain in 74% of cases. MPS is a treatable condition that can be well managed by manual, needling, and injection techniques. However, the syndrome is often undiagnosed by physicians. This is due likely to a lack of knowledge of this disorder and especially a lack of knowledge and tools regarding treatment. Regarding the high prevalence of MPS found in the present study, we note that all five participating PCP had thorough education and knowledge of the syndrome; thus possibly displaying a bias toward this diagnosis. In addition, it is possible that the cases referred to these PCPs were a non-random sample of pain patients, but those with a preponderance for myofascial pain. For example, patients with ongoing cancer therapy could be treated by their oncologists and pain specialists in a tertiary pain setting and were not part of our sample.

Pain is a multifactorial experience associated with physical, psychological, and emotional factors that play important roles in patients' disability and quality of life. The participating PCPs are trained in the biopsychosocial approach, and although not investigated in the current work, this approach was inherent in the treatment of all patients. In all the pain management clinics involved in this study, dry needling was the most common treatment for patients with pain complaints. This is evidence-based safe and effective technique that is widely accepted and used in Israel and in other countries to treat MPS (Halle and Halle 2016; Kietrys et al. 2013; Henschke et al. 2010). The method is based on inserting needles into the muscles involved, to relieve the intensity of pain and improve the range of motion. Trained instructors of the Israeli Society for Musculoskeletal Medicine offer courses in dry needling to family practitioners. This is important due to a high prevalence of MPS and the ease of administration of dry needling in the primary care setting.

A small proportion of patients in our clinics were referred to imaging studies for evaluation of their pain problems. Some of the patients were referred by physicians from other disciplines such as orthopedic surgeons, rheumatologists, and family practitioners; and already had imaging studies performed upon entry into the current study. Even so, the physicians participating in this study used accepted guidelines for further referral to imaging; the rate of that referral was low. Thus, it would seem that with thorough medical history taking and physical examination, most pain problems can be diagnosed in a community setting, without further imaging.

This study was performed in five community based clinics in Israel, in different settings (city, town, and rural-based clinics), managed by three health maintenance organizations. The population was heterogeneous, yet all patients were entitled to uniform primary care medical services according to the Israeli health care basket. The diagnoses reached, treatments offered, and clinical outcomes were remarkably similar, despite significant differences in patient demographics. Nonetheless, lack of a control group is a limitation of this observational study. The paucity of pain consultation services is striking. In Israel, only 80 physicians are board-certified in pain medicine, 22 residents and approximately 20 other physicians work predominantly in the field of pain medicine (data from the Israel Pain Association). In a country of 8.8 m citizens, there is thus approximately one pain specialist per 100,000 population. Patients who suffer from pain are seen primarily by family physicians. Since most of these physicians did not receive sufficient training in pain treatment, either as students or as residents, their knowledge in the field is based primarily on postgraduate education, which is often deficient. During recent years, educational programs have provided training to PCPs in pain management; yet the number of participants is still small.

Secondary care physicians in pain medicine are certified in CME programs in pain and musculoskeletal medicine, having gained extensive knowledge and proficiency in the field. They are able to manage a large portion of patients suffering from pain in the community, especially those with more complex and persistent pain problems. Shifting the management of chronic pain patients from tertiary centers to the community, including referrals to secondary pain clinics, will necessitate a profound change in the training of physicians at various levels of treatment. The efficacy of this shift must be evaluated with measurable parameters. In the biological realm, this includes such parameters as visual analogue scale, the use of pain medication, days off from work and physician visitation rates. In the psychosocical realm, this includes the measures of quality of life and disability .

In summary, cost effective strategies that improve the management of pain in the primary care setting are urgently needed. In this paper we described a novel patient-centered pain care model by which certified physicians serve as secondary referral addresses, mid-way between primary care medicine and tertiary pain centers. We believe that such model holds promise for the future of pain management in primary care. More studies are needed to examine the pain improvement outcomes of this strategy.

Acknowledgments We thank Dr. Alon Margalit and Dr. Aviva Elad for participation and data collection for this study.

Conflicts of Interests The authors declare no conflicts of interest in relation to this article.

References

Breivik H, Collett B, Ventafridda V, Cohen R, Gallacher D (2006) Survey of chronic pain in Europe: prevalence, impact on daily life, and treatment. Eur J Pain 10:287–333

Carmeli E (2014) The invisibles: unpaid caregivers of the elderly. Front Public Health 2:91

Declaration of Montréal (2017) http://www.iasp-Pain.org/DeclarationofMontreal?navItemNumber=582. Accessed 25 Jan 2017

Dubois MY, Gallagher RM, Lippe PM (2009) Pain medicine position paper. Pain Med 10:972–1000

Dueñas M, Ojeda B, Salazar A, Mico JA, Failde I (2016) A review of chronic pain impact on patients, their social environment and the health care system. J Pain Res 9:457–467

Fishbain DA, Goldberg M, Meagher BR, Steele R, Rosomoff H (1986) Male and female chronic pain patients categorized by DSM-III psychiatric diagnostic criteria. Pain 26:181–197

Fleckenstein J, Zaps D, Rüger LJ, Lehmeyer L, Freiberg F, Lang PM, Irnich D (2010) Discrepancy between prevalence and perceived effectiveness of treatment methods in myofascial pain syndrome: results of a cross-sectional, nationwide survey. BMC Musculoskelet Disord 11:32

Fogelman Y, Kent J (2015) Efficacy of dry needling for treatment of myofascial pain syndrome. J Back Musculoskelet Rehabil 28:173–179

Freedman KB, Bernstein J (2002) Educational deficiencies in musculoskeletal medicine. J Bone Joint Surg Am 84:604–608

Gerwin RD (1995) A study of 96 subjects examined both for fibromyalgia and myofascial pain. J Musculoskelet Pain 3:121

Gureje O, Gregory ES, Von Korff M (2001) Cross-national study of the course of persistent pain in primary care. Pain 92:195–200

Halle JS, Halle RJ (2016) Pertinent dry needling considerations for minimizing adverse effects- part two. Int J Sports Phys Ther 11:810–819

Henschke N, Ostelo RW, van Tulder MW, Vlaeyen JW, Morley S, Assendelft WJ, Main CJ (2010) Behavioural treatment for chronic low-back pain. Cochrane Database Syst Rev 7:CD002014

Kalichman L, Vulfsons S (2010) Dry needling in the management of musculoskeletal pain. Am Board Fam Med 23:640–646

Kietrys DM, Palombaro KM, Azzaretto E, Hubler R, Schaller B, Schlussel JM, Tucker M (2013) Effectiveness of dry needling for upper-quarter myofascial pain: a systematic review and meta-analysis. J Orthop Sports Phys Ther 43:620–634

Kroenke K (2014) A practical and evidence-based approach to common symptoms: a narrative review. Ann Intern Med 161:579–586

Leverence RR, Williams RL, Potter M, Fernald D, Unverzagt M, Pace W, Parnes B, Daniels E, Skipper B, Volk RJ, Brown AE, Rhyne RL, Net Clinicians PRIME (2011) Chronic non-cancer pain: a siren for primary care -a report from the PRImary care MultiEthnic network (PRIME net). J Am Board Fam Med 24:551–561

Lincoln LE, Pellico L, Kerns R, Anderson D (2013) Barriers and facilitators to chronic non-cancer pain management in primary care: a qualitative analysis of primary care providers' experiences and attitudes. J Palliative Care Med S3:001

Mäntyselkä P, Kumpusalo E, Ahonen R, Kumpusalo A, Kauhanen J, Viinamäki H, Halonen P, Takala J (2001) Pain as a reason to visit the doctor: a study in Finnish primary health care. Pain 89:175–180

Matzkin E, Smith EL, Freccero D, Richardson AB (2005) Adequacy of education in musculoskeletal medicine. J Bone Joint Surg Am 87:310–314

Mehta N, Inturrisi CE, Horn SD, Witkin LR (2016) Using chronic pain outcomes data to improve outcomes. Anesthesiol Clin 34:395–408

Minerbi A, Vulfsons S (2013) Pain medicine in crisis – a possible model toward a solution: empowering community medicine to treat chronic pain. Rambam Maimonides Med J4:e0027

Peppin JF, Cheatle MD, Kirsh KL, McCarberg BH (2015) The complexity model: a novel approach to improve chronic pain care. Pain Med 16:653–666

Skootsky SA, Jaeger B, Oye RK (1989) Prevalence of myofascial pain in general internal medicine practice. West J Med 151:157–160

Speerin R, Slater H, Li L, Moore K, Chan M, Dreinhöfer K, Ebeling PR, Willcock S, Briggs AM (2014) Moving from evidence to practice: models of care for the prevention and management of musculoskeletal conditions. Best Pract Res Clin Rheumatol 28:479–515

Steglitz J, Buscemi J, Ferguson MJ (2012) The future of pain research, education, and treatment: a summary of the IOM report "relieving pain in America: a blueprint for transforming prevention, care, education, and research". Transl Behav Med 2:6–8

Thomsen AB, Sørensen J, Sjøgren P, Eriksen J (2002) Chronic non-malignant pain patients and health economic consequences. Eur J Pain 6:341–352

Turk DC, Fillingim RB, Ohrbach R, Patel KV (2016) Assessment of psychosocial and functional impact of chronic pain. J Pain 17(Suppl):T21–T49

VHA Directive 2009-053 (2009) Pain Management. Department of Veterans Affairs, Washington, DC

WHO (2003) The burden of musculoskeletal conditions at the start of the new millennium: report of a WHO scientific group. WHO, Geneva, Switzerland

Advs Exp. Medicine, Biology - Neuroscience and Respiration (2018) 38: 99–100
https://doi.org/10.1007/978-3-319-74080-5

Index